PfMP® EXAM
PRACTICE TESTS
AND STUDY GUIDE

Best Practices and Advances in Program Management Series

Series Editor
Ginger Levin

RECENTLY PUBLISHED TITLES

PfMP® Exam Practice Tests and Study Guide
Ginger Levin

Program Management Leadership: Creating Successful Team Dynamics
Mark C. Bojeun

Successful Program Management: Complexity Theory, Communication, and Leadership
Wanda Curlee and Robert Lee Gordon

From Projects to Programs: A Project Manager's Journey
Samir Penkar

Sustainable Program Management
Gregory T. Haugan

Leading Virtual Project Teams: Adapting Leadership Theories and Communications Techniques to 21st Century Organizations
Margaret R. Lee

Applying Guiding Principles of Effective Program Delivery
Kerry R. Wills

Construction Program Management
Joseph Delaney

Implementing Program Management: Templates and Forms Aligned with the Standard for Program Management, Third Edition (2013) *and Other Best Practices*
Ginger Levin and Allen R. Green

Program Management: A Life Cycle Approach
Ginger Levin

Program Management Leadership: Creating Successful Team Dynamics
Mark C. Bojeun

The Essential Program Management Office
Gary Hamilton

Project Management in Extreme Situations: Lessons from Polar Expeditions, Military and Rescue Operations. and Wilderness Explorations
Monique Aubry and Pascal Lievre

Program Management for Business: Aligning and Integrating Strategy, Process, Technology, People and Measurement
Satish P Subramanian

PfMP® EXAM PRACTICE TESTS AND STUDY GUIDE

DR. GINGER LEVIN, PMP, PgMP

CRC Press
Taylor & Francis Group
Boca Raton London New York

CRC Press is an imprint of the
Taylor & Francis Group, an **informa** business

AN AUERBACH BOOK

Parts of A Guide to the Project Management Body of Knowledge, 2014, are reprinted with permission of the Project Management Institute, Inc., Four Campus Boulevard, Newtown Square, Pennsylvania 19073-3299 U.S.A., a worldwide organization advancing the state of the art in project management.

Parts of The Standard for Portfolio Management, 2014, are reprinted with permission of the Project Management Institute, Inc., Four Campus Boulevard, Newtown Square, Pennsylvania 19073-3299 U.S.A., a worldwide organization advancing the state of the art in project management.

"OPM3" is a trademark of the Project Management Institute, Inc., which is registered in the United States and other nations.

"PfMP" is a certification mark of the Project Management Institute, Inc., which is registered in the United States and other nations.

"PMBOK" is a trademark of the Project Management Institute, Inc., which is registered in the United States and other nations.

"PMI" is a service and trademark of the Project Management Institute, Inc., which is registered in the United States and other nations.

"PMP" is a certification mark of the Project Management Institute, Inc., which is registered in the United States and other nations.

CRC Press
Taylor & Francis Group
6000 Broken Sound Parkway NW, Suite 300
Boca Raton, FL 33487-2742

First issued in hardback 2017

© 2014 by Taylor & Francis Group, LLC
CRC Press is an imprint of Taylor & Francis Group, an Informa business

No claim to original U.S. Government works

ISBN 13: 978-1-138-44028-9 (hbk)
ISBN 13: 978-1-4822-5100-5 (pbk)

Contents

Preface

Based on experience in helping people to prepare for the PMP® and the PgMP® Exams, we know that you will have questions, such as "What topics are covered on the exam?" and "What are the questions like?" Not surprisingly, some of the most sought-after study aids are practice tests, which are helpful in two ways (1) taking practice tests increases your knowledge of the kinds of questions, phrases, terminology, and sentence construction that you will encounter on the real exam; and (2) taking practice tests provides an opportunity for highly concentrated study by exposing you to a breadth of portfolio management content generally not found in a single reference source.

We developed this specialty publication with one simple goal in mind—that is, to help you study for and pass the PfMP® certification exam. Because the Project Management Institute (PMI®) does not sell past exams for prospective certification purposes, the best option is to develop practice test questions that are as representative as possible. And that is exactly what is in this book.

This book contains study hints, a list of exam topics, and multiple-choice questions for each of the five domains covered in the PfMP® exam, according to the *Portfolio Management Professional (PfMP®) Examination Content Outline (2013)*. We have prepared 20 practice questions in each of these five domains. We have also included two 170-question representative practice tests.

As has been done in our other Exam Practice Test & Study Guide books, we have included a plainly written rationale for each correct answer along with a supporting reference list. Each question also shows the applicable task from the *Examination Content Outline* (ECO). References are provided at the end of this study guide for the five domains covered in the exam: Strategic Alignment, Governance, Portfolio Performance, Portfolio Risk Management, and Communications Management.

Many questions are scenario based, as are those on the PfMP® exam. While some are definition questions, they are often based on a situation.

For those who speak English as a second language (ESL), each question and answer in the practice tests are written in a way that words, terms, or phrases that could be confusing to people who are not fluent in English are omitted. Although the language issue may concern you, and rightfully so, the only difference between you and those who speak English as their first language is the amount of time it takes to complete the exam. If you can grasp the content expressed in this publication, then we believe that a few colloquialisms or ambiguous terms on the real exam ultimately will not determine whether you pass or fail. Your subject matter knowledge will do that.

Earning the PfMP® certification is a prestigious accomplishment, but studying for it should not be difficult if you use the tools available.

Good luck on the exam!

Dr. Ginger Levin, PMP, PgMP
Lighthouse Point, Florida

Acknowledgments

We want to acknowledge the efforts of our publisher, CRC Press, and especially that of Mr. John Wyzalek. The CRC team, who include Christopher Manion, Jessica Vakili, and Randy Burling, worked tirelessly to publish this book so it would be available according to the release of the new PfMP® exam from PMI®.

Special thanks as well to Dr. David Hilson and Dr. Werner Meyer for their assistance in answering several questions concerning terms in the Portfolio Risk Management domain.

About the Author

Dr. Ginger Levin is a senior consultant and educator in portfolio, program and project management with more than 45 years experience in the public and private sectors. Her specialty areas include portfolio management, program management, business development, maturity assessments, metrics, organizational change, knowledge management, and the project management office. She is active in providing training to others as they study for their PfMPs® and Program Management Professional (PgMPs®). She is an Adjunct Professor for the University of Wisconsin-Platteville in its master's degree program in project management and at SKEMA, France in its doctoral programs in project management. Before her consulting and teaching career, she was President of GLH, Incorporated, a woman-owned small business in the Washigton, DC area for 15 years, specializing in project management. Earlier, she had a career in the U.S. Government, working for six agencies in positions of increasing responsibility for 14 years. As an intern with the Southern Railway, she was fortunate to work in what today would be known as a Portfolio Management Office, where she learned how to best make and present business cases for new products and services and also how to work to prioritize components. In her consulting work, she has had the opportunity to implement portfolio management processes and systems for several public and private sector organizations.

Dr. Levin is the editor of *Program Management A Life Cycle Approach*, the author of *Interpersonal Skills for Portfolio, Program, and Project Managers* and the coauthor of *Program Management Complexity: A Competency Model, Implementing Program Management: Templates and Forms Aligned with the Standard for Program Management*—Third Edition (2013) and a similar book in 2008; *Project Portfolio Management; Metrics for Project Management; Essential People Skills for Project Managers; Achieving Project Management Success Using Virtual Teams; The Advanced Project Management Office: A Comprehensive Look at Function and Implementation; People Skills for Project Managers;* and the *Business Development Capability Maturity Model.* Since 1996 she has been active as a co-author in ESI International's *PMP® Challenge!* (now in its sixth edition) and *PMP® Prep Practice Test Study Guide* (now in its ninth edition). Similarly, she is the co-author of the *PgMP® Challenge!* and the *PgMP® Prep Practice Test and Study Guide* (now in its fourth edition).

Dr. Levin is a member of the Project Management Institute (PMI®) and a frequent speaker at PMI® Congresses and Chapters and the International Project Management Association. She is certified by PMI® as a Project Management Professional (PMP®), and a Program Management Professional (PgMP®), and she was the second person in the U.S. to earn the PgMP® designation. She is also certified as an Organizational Project Management Maturity Model (*OPM3*®) Professional.

Dr. Levin holds a doctorate in public administration from The George Washington University, where she also received the Outstanding Dissertation Award for her research on large organizations. She also holds a master of science in business administration from The George Washington University and a bachelor of business administration from Wake Forest University.

Introduction

The Program Management Professional (PfMP®) Exam: Practice Test & Study Guide includes five sections, each of which corresponds to one of the five domains described in the *Portfolio Management Professional (PfMP®) Examination Content Outline* (2013). Each section contains study hints, a list of major topics that may be encountered on the exam, and 20 multiple-choice practice questions complete with an answer sheet, an answer key that includes a rationale for each correct answer, a link to the corresponding task in the ECO, and a bibliographic reference for further study if needed.

We have also included two complete practice tests, each consisting of 170 questions,[1] that follow the blueprint of the real PfMP® exam as described in the ECO. For example—

- 25 percent of the questions relate to strategic alignment
- 20 percent relate to governance
- 25 percent relate to portfolio performance
- 15 percent relate to portfolio risk management
- 15 percent relate to communications management

To use this study guide effectively, first read the *Standard for Portfolio Management* – Third Edition (2014) and the ECO. Next try one of our Practice Tests on line to see your areas of strengths and weaknesses. Then, go through this book on one section at a time because this order reflects the *Examination Content Outline*. Start by reading the study hints, which provide useful background on the content of the PfMP® exam and identify the emphasis placed on various topics. Familiarize yourself with the major topics listed. Then, answer the practice questions that follow, recording your answers on the answer sheet provided. Finally, compare your answers to those in the answer key. The rationales provided should clarify any misconceptions. For further study and clarification, consult the ECO and bibliographic references. Since the ECO is the "blueprint" for the exam, each question is linked to specific tasks in each domain in the ECO, which means the references from the *Standard* at times may be in different knowledge areas.

After you have finished answering the questions that follow each section, it is time to take the practice tests. Note your answers to the first Practice Test you took on line on the sheets provided, compare your answers to the answer key, and use your results to determine what areas you need to study further.[2] Then, take the other Practice Test on line and later in the book.

To make the most of this specialty publication, use it regularly. The tips listed below will help you to get the most out of your preparation for the PfMP® exam:

■ Take and retake the practice tests. Photocopy the answer sheets so that you have a clean one to use each time you retake the practice tests.
■ Consider convening a study group to compare and discuss your answers with those of your colleagues. This method of study is a powerful one. You will learn more from your colleagues than you ever thought possible.
■ Make sure that you have a solid understanding of the exam topics provided in each section.
■ Consult our bibliography, or other sources that you have found useful, for further independent study.
■ Most importantly, create a study plan and stick to it. Your chances of success increase dramatically when you set a goal and dedicate yourself to meeting it.

To further enhance your study, **the two Practice Tests are available on line at: http://www.ittoday.info/pfmp/examhome.html**. To access the tests, use this following passcode: K23799. This web-based version of these exams includes a clock so you can see how well you do in the allotted four hours for the exam. You will see your answers according to the number of questions you answered correctly by domain. Also, you will see your answers according to a proprietary algorithm, designed by the author, as to whether you were Proficient, Moderately Proficient, or Below Proficient.

In either the print or on the web, the practice tests are an essential study tool that was created with one goal in mind: helping you to pass your exam and become PfMP® certified.

List of Acronyms

ANCI: Annual Net Cash Inflows
AP: Asia Pacific
CEO: Chief Executive Officer
CFO: Chief Financial Officer
CIO: Chief Information Officer
CMMI: Capability Maturity Model Integration
ECO: Examination Content Outline
EMV: Expected Monetary Value
EPMO: Enterprise Project, Program or Portfolio Management Office
ERP: Enterprise Resource Planning
ESL: English as a Second Language
FDA: Food and Drug Administration
HAACP: Hazard Analysis and Critical Control Point
HVAC: Heating, Ventilating, and Air Conditioning
IPO: Initial Public Offering
IRR: Internal Rate of Return
IS: Information Systems
IT: Information Technology
KPIs: Key Performance Indicators
NDA: Non-disclosure Agreements
NPV: Net Present Value
OPM: Organizational Project Management
***OPM3*®:** Organizational Project Management Maturity
PfMP®: Portfolio Management Professional
PgMP®: Program Management Professional
PMBOK® Guide: *A Guide to the Project Management Body of Knowledge*
PMI®: Project Management Institute
PMIS: Portfolio Management Information System
PMO: Project, Program or Portfolio Management Office
RACI: Responsible, Accountable, Consulted, Informed
RBS: Risk Breakdown Structure
REIT: Real Estate Investment Trust
RFP: Request for Proposals

ROI: Return on Investment
SMART: Specific, Measurable, Attainable, Realistic, and Time Based
SMEs: Subject Matter Experts
SWOT: Strengths, Weaknesses, Opportunities, Threats
WBS: Work Breakdown Structure

Strategic Alignment

Study Hints

The Strategic Alignment management questions on the PfMP® certification exam, which constitute 25% of the exam or 43 questions.

These questions are part of the Portfolio Strategic Management Knowledge area as well as in the Defining and Aligning Process Groups in *The Standard for Portfolio Management*—Third Edition (2013), referred to as *The Standard* throughout this book. There are four processes in Portfolio Strategic Alignment:

1. Develop Portfolio Strategic Plan
2. Develop Portfolio Charter
3. Define Portfolio Roadmap
4. Manage Strategic Change

The first three processes are from the Defining Process Group, while Manage Strategic Change Process is from the Aligning Process Group.

All benefit from the existing organizational process assets, portfolio process assets, portfolio reports, and enterprise environmental factors, which are common inputs and outputs to most processes. They are often updated as the processes are performed, and you should recognize how they are used in each process, as well as how the process leads to specific updates to them

Since this area contains three major deliverables in portfolio management, it is essential to know their contents and the best practices to follow to develop them and to ensure these deliverables are aligned to the strategy of the organization. Further as the organization's strategy changes, these documents plus other organizational and portfolio process assets require change and updates.

As the Portfolio Strategic Plan is developed, an inventory of existing portfolio components is used along with portfolio strategic alignment. Review Figure 4-4 from *The Standard* for an integrated view of the strategy based on a high-level timeline. Also as the plan is developed, prioritization analysis is used as it provides an approach as to whether new components (programs, projects, or operational work) should be added to the portfolio, existing components continued, or existing components terminated, as priorities and goals may have changed, and the resources from them can be allocated more efficiently on other components.

Various prioritization models can be used ranging from ones that are simple to ones that are sophisticated. The objective is to use the model to determine a score for each component. Establishing this model is more difficult than it seems as everyone in the organization basically is a stakeholder and must follow the criteria that are established for scoring. The strategic plan then ensures the vision and objectives of the portfolio are aligned with that of the organization.

The Portfolio Charter also is developed in this process, and it is comparable to that of a charter for a program or project manager as it authorizes the portfolio manager to assign resources to components and execute the various portfolio processes. It builds on the portfolio strategic plan, and in developing it, scenario analysis may be helpful as it can show different combinations of components to help determine the most useful structure of the portfolio. Capability and capacity analysis cannot be overlooked as in most organizations there are more components, or potential components, than resources that are available to execute them. Such analyses enable the organization to determine constraints and how best to handle them. As the charter is developed, it may cause a need to revisit the portfolio strategic plan to see if it requires updates.

Additionally, the portfolio roadmap is developed. It is an expansion of a roadmap at the program level as it shows all the components in the portfolio needed to achieve the organization's overall strategy in a chronological way. As it serves as a high-level plan, it also focuses on dependencies and shows component alignment with strategies, maps milestones, and shows challenges and risks. It is an excellent communications tool that changes frequently. In developing it, the portfolio strategic plan, the charter, and the portfolio are used. Analyses focus on interdependencies between components, cost-benefit, and prioritization. Review Figure 4-8 in *The Standard* to see an example of a roadmap.

Managing strategic change is critical as portfolio management responds to these changes to ensure the portfolio's components remain in alignment or to determine if new components need to be added, or others removed. It is a dynamic process as the portfolio is not static. It is part of the Aligning Process Group with an emphasis on the portfolio manager working to manage changes in strategy at the organizational level that affect the portfolio and to accept or act on any changes that then effect portfolio planning. These changes are expected, and to assist the portfolio manager, he or she uses the portfolio strategic plan, charter, roadmap, and the portfolio management plan [developed in the governance process to document the intended approach to manage the portfolio]. Tools and techniques include stakeholder analysis to meet stakeholder expectations as the portfolio changes, gap analysis to assess the current state with the vision or 'to be' state at the organizational level, and readiness analysis to determine what must be done to bridge any gaps that exist. This process results in updates to documents and portfolio process assets.

Following is a list of the major topics in the Strategic Alignment domain. Use this list to focus your study efforts on the areas that are most likely to appear on the exam.

Major Topics

Portfolio definition
Portfolio management definition
Relationships between—

■ Portfolios, programs, projects, and operations
■ Portfolio management and organizational strategy
■ Portfolio management and organizational project management
■ Portfolio management and the PMO (project, program, or portfolio management office)

Impact of portfolio management on organizational strategy and objectives

■ Maintaining portfolio alignment
■ Allocating financial, human, material, and equipment resources
■ Measuring component performance
■ Managing risks
■ Adding, deferring, or deleting components
■ Defining the portfolio vision and plan

Business Value
Role of the portfolio manager and key areas of expertise and skills

■ Organizational strategy
■ Organizational criteria for portfolio management
■ Recognizing all the portfolio components
■ Following agreed-upon processes

Assessing the current portfolio management process

■ Assessment activities
■ Assessment results

Defining Process Group
Aligning Process Group
Common inputs and outputs

■ Portfolio process assets
■ Portfolio reports
■ Organizational process assets
■ Enterprise environmental factors

Purpose of portfolio strategic management

Develop Portfolio Strategic Plan process

■ Purpose
■ Inputs
  – Organizational strategy and objectives
    • Long-term direction
    • Vision
    • Goals
    • Objectives
  – Inventory of work
  – Portfolio process assets
    • Plans
    • Policies
    • Procedures
    • Guidelines
  – Organizational process assets
    • Strategy and objectives
    • Mission and vision
    • Prioritization
    • Resources
  – Enterprise environmental factors
    • Organizational structure
    • Stakeholder risk tolerances
    • Market conditions
    • Human resources
■ Tools and Techniques
  – Portfolio component inventory
    • Initial portfolio
  – Strategic alignment analysis
    • New or changing organizational strategy
    • Gaps in focus, investment, or alignment
    • High-level timeline
  – Prioritization analysis
    • Model to guide decisions regarding components
    • Simple scoring model to complex model
■ Outputs
  – Portfolio strategic plan
    • Purpose
    • Vision and objectives
    • Organizational structure
    • Measurable goals and guidelines
    • Allocation of funds
    • Benefits, performance results, value expected
    • Needed communications

- Assumptions, constraints, dependencies, risks
- Type and quantity of resources
- Prioritization model
  – Portfolio
    - List of components

Develop Portfolio Charter

- ■ Purpose
- ■ Inputs
  – Portfolio strategic plan
    - Vision and objectives
    - Expected benefits
    - Key risks, dependencies, constraints
  – Portfolio process assets
    - Plans, policies, procedures, guidelines
    - Stakeholder relationships
    - Scope, benefits, portfolio goals
  – Enterprise environmental factors
    - Corporate accounting structure
    - Functional structure
- ■ Tools and techniques
  – Scenario analysis
    - Evaluate outcomes based on alternatives
  – Capability and capacity analysis
    - Resource availability—internal and external
    - Skill set limitations
    - Financial constraints
    - Asset constraints
- ■ Outputs
  – Updates
    - Portfolio strategic plan
    - Portfolio process assets
  – Portfolio charter
    - Purpose
    - Authority of the portfolio manager
    - Portfolio structure
    - Delivering value
    - Objectives, justification, sponsors, roles and responsibilities
    - Stakeholders and their expectations and requirements
    - Communication requirements
    - High-level scope, benefits, critical success criteria, resources
    - High-level time frame
    - Assumptions, constraints, dependencies, risks

Define Portfolio Roadmap

■ Purpose
■ Inputs
  – Portfolio strategic plan
    • Goals, objectives, and strategies
    • Prioritization guidelines
  – Charter
    • Structure
    • Scope
    • Constraints
    • Dependencies
    • Resources
  – Portfolio
    • Prioritization
    • Dependencies
    • Organization areas
■ Tools and techniques
  – Interdependency analysis
    • Dependencies
    • Participants in the analysis
  – Cost-benefit analysis
    • Costs and benefits
    • Qualitative considerations
  – Prioritization analysis
    • Compare strategic objectives
    • Prioritize objectives
    • Perform strategic assessment
■ Outputs
  – Portfolio roadmap
    • Purpose
    • High-level strategic direction and information
    • Chronological view

Manage Strategic Change

■ Purpose
■ Inputs
  – Portfolio strategic plan
    • Current components
    • Highest strategic value
  – Charter
    • Alignment with portfolio

- Portfolio
  - Component mix and strategic direction
- Roadmap
  - Evaluate dependencies
  - 'As is' and 'to be' state
- Portfolio management plan
  - Intended approach to manage the portfolio
- Portfolio process assets
  - Analysis and assessment tools and techniques
■ Tools and techniques
  - Stakeholder analysis
    - Align expectations with changes
    - Interviews
    - Determine expectations and pain points
    - Risk tolerances
    - Problems
    - Change impacts and issues
  - Gap analysis
    - Compare component mix with new strategic direction
  - Readiness assessment
    - Determine if, when, what, and how to implement change
■ Outputs
  - Updates to
    - Portfolio strategic plan
    - Charter
    - Portfolio
    - Roadmap
    - Portfolio management plan
    - Portfolio process assets

Practice Questions

INSTRUCTIONS: Note the most suitable answer for each multiple-choice question in the appropriate space on the answer sheet.

1. Working as a portfolio manager in your food company, you are responsible for the food additive portfolio. Each new food additive requires approval from the Food and Drug Administration (FDA). Recently, the company built a new plant to be able to manufacture chocolate that would not require any sugar, but consumers would not notice any differences in flavor. A FDA inspector arrived at the plant before the product was produced and found one of the manufacturing lines was not working as planned. This now delays production of the new food additive, and another inspection will be needed. In such a situation, the portfolio manager must:

   a. Recommend that work on this new product be deferred
   b. Ensure compliance with regulatory requirements
   c. Work closely with the Quality Assurance Department to determine how best to satisfy FDA's requirements
   d. Recommend the next steps to the Portfolio Review Board working in conjunction with quality management

2. Assume you have been asked to assess whether there is a need to revise the portfolio components. This request may be a result of:

   a. New leadership
   b. Opportunities to be pursued
   c. A change in the risk tolerance levels
   d. The need to enhance ROI

3. Organizational performance can be characterized in several different ways such as taking a goal approach, a system's resource approach, or a constituency approach. Each organization uses different ways to describe it, but in terms of portfolio management, it:

   a. Serves as the basis for the portfolio management plan
   b. Is documented in the portfolio strategic plan
   c. Forms the governance structure
   d. Is described in the portfolio charter

4. As your University suffered a 55% budget cut, not all of the components in the portfolio can be continued. Plus, some proposed components will not be able to be considered. As the portfolio manager, you set up criteria by which to rank the existing and proposed components, and you have maintained an accurate inventory of the work in portfolio. Your criteria use a factor of 0 being of least important to 10, the most important. The following is an example.

Criteria	Item	Score	Average Criterion
Financial	Net Present Value	8	5
	Payback Period	5	
	Internal Rate of Return	2	
Customer Satisfaction	Enhanced Benefits	7	8
	Improved Customer Relationship Management	9	
Strategic Goals	Alignment with the University's Strategic Goals	7	7
Competitive Advantage	Attraction to New Students	6	6.3
	Enhanced Growth	9	
	Attraction to New Faculty	4	
Technical Merit	Probability of Success	9	9
	Complexity	9	
Total Score			35.3 or 88%

This is an example of a scoring model that is:

a. A simple sum of selected criteria
b. One that would take time to use across all portfolio components
c. One that requires agreement among stakeholders as to the ranking process
d. Maximizes the business value

5. Various components comprise a portfolio. Assume you are working as a portfolio manager for a business unit in your manufacturing company. There is one characteristic that applies to all of the components in the portfolio in your business unit, which is:

a. These components are related in some way
b. The components once they are in the portfolio remain in the portfolio
c. The components can be quantified
d. The components are determined based on contributions to business benefits

6. Capability and capacity analysis are two techniques that complement one another to help determine the portfolio structure as there may be limiting factors for the number of components an organization can execute. These limiting factors include:

   a. Capability factors such as resource availability
   b. Capability factors such as lack of key skill sets
   c. Capacity factors such as financial and asset considerations
   d. Capability factors such as financial and asset considerations

7. Your asphalt company recognizes that several of its products are obsolete, and it also realizes it is losing market share as a result. It also has outdated practices, and it is unclear as to whether resources are allocated effectively. Your job as the portfolio manager is to ensure resources are allocated to those components that will provide the most business value. This means as a first step you should focus on:

   a. Establishing competency profiles of staff member expertise
   b. Creating an up-to-date list of components
   c. Determining based on market analysis new products for the company to pursue
   d. Evaluating financial capacity for new components and additional resources

8. Multiple weighted criteria are useful ranking approaches especially since in most organizations greater attention is given to certain components than to others. Consider the following example and assume only one component can be added to the portfolio when the Portfolio Review Board meets.

Criteria	Weight	Project A		Project B		Program A		Operational Activity	
		Points	Weighted Points	Points	Weighted Points	Points	Weighted Points	Points	Weighted Points
ROI	3	4		5		3		4	
New Business	4	2		3		5		3	
Potential Savings	3	5		4		2		4	
First to Market	5	4		3		5		4	
Total		15		15		15		15	

Your recommendation is to select:

a. Project A
b. Project B
c. Program A
d. Operational Activity

9. Assume you are leading a team to recommend the factors to use to consider the appropriate mix of components in the organization's portfolio as it implements a prioritization process. You are to present your recommendations to the Executive Committee on Friday. One factor you and your team consider to be especially important is:

a. Complexity
b. Risk versus reward
c. Customer satisfaction
d. Extent of organizational change

10. Recently, your city declared bankruptcy. As the portfolio manager working for the Mayor, you knew this approach was needed given the declining economy, the debt it had occurred, and its inability to raise money through the sale of bonds. To recover, drastic strategic changes are needed. As the portfolio manager, you are conducting interviews and holding focus groups with stakeholders, including City employees, citizens, and consumer and social activist groups. As you do so, you find:

    a. It is helpful to group stakeholders by category
    b. Many were not prepared for this change
    c. There is a loss of confidence in the executive leadership team
    d. There are only a limited number of people who truly understand the significance of this bankruptcy

11. With management by projects and programs the norm and not the exception in many organizations, assume the Board of Directors of your cereal company decided to become a project-based organization two years ago and not work in functional silos. This approach was a cultural change that has taken time to fully embrace, and a portfolio management committee also has been set up. To assess its progress it is conducting an Organizational Project Management Maturity (*OPM3*) assessment. One reason to do so is to:

    a. Determine the organization's overall level of maturity in project, program, and portfolio management
    b. Focus on continuous improvement
    c. Ensure processes in place are repeatable and easy to follow
    d. Link project, program, and portfolio principles with organizational enablers

12. Assume you are striving to create an up-to-date list of all approved portfolio components to have an organized portfolio your giant aerospace company can use for ongoing evaluation, selection, and prioritization. Resources are limited, and must be allocated to those programs and projects that will provide the greatest business value. To create such a list a best practice is to:

    a. Follow the portfolio performance plan
    b. Review organizational process assets
    c. Select evaluation criteria
    d. Review the portfolio charter

13. While your organization has been doing portfolio management by following a standard process for several years, it has not prepared a portfolio strategic plan. Assume you were asked to prepare one by the Portfolio Review Board. First, you should:

    a. Review the organization's mission statement
    b. Determine the values the organization follows in decision making
    c. Conduct a strategic alignment assessment
    d. Conduct a capability and capacity analysis

14. Your IT Company is determined to change its image from one that delivers services late, over budget, with the need for rework as it has been blamed for failure to deliver successfully a government-required system. It is instead going to focus on providing external PMO support and will slowly complete its remaining IT products. Assume you are leading this strategic change since you are the portfolio manager. You want to determine how best to implement this change so you decide to:

    a. Conduct a gap analysis
    b. Conduct a readiness analysis
    c. Assess the risks for each factor that may impede the change and determine a mitigation strategy
    d. Perform a strategic assessment against the existing portfolio

15. Your company is determined to be the market leader in TVs that combine the best features of LCD and Plasma and enable video streaming without the need to use a computer, tablet, or smart phone. To succeed, your Board of Directors authorized your CEO to acquire other companies that offer similar types of TVs, and through these acquisitions your company also acquires intellectual property to ensure the advanced technology can be developed. Thus far, three companies have been acquired with this strategic change. As the portfolio manager, you need to update the portfolio management plan because of changes in the:

    a. Benefits
    b. Constraints
    c. New strategic objectives
    d. Management approach

16. One way to show the high-level strategic direction of the portfolio is to prepare a:

    a. Communications management plan
    b. Master schedule
    c. Portfolio roadmap
    d. Portfolio management plan

17. The roadmap forms the initial basis to establish dependencies, which are:

    a. Part of each component
    b. Especially important for standalone projects to show why they are being pursued
    c. Within the portfolio and between organizational areas
    d. The basis for prioritization mapping

18. As you work to implement portfolio management in your consulting firm, which has lacked it since it was established ten years ago, you recognize a prioritization model is essential. The consulting firm's approach was to focus on win rate rather than capture ratio, and it typically lacks needed resources to handle the work and has been unable to focus on customer relationship management. This prioritization model is:

    a. Part of the portfolio process assets
    b. Contained in the portfolio strategic plan
    c. Contained in the portfolio management plan
    d. Part of the portfolio structure

19. It is a best practice for the portfolio manager as he or she prepares the portfolio strategic plan to:

    a. Involve stakeholders at all levels in the process
    b. Communicate its importance throughout the organization
    c. Integrate and respond to changes in the portfolio
    d. Ensure funds are available to allocate to each category in the portfolio

20. An example of a simple prioritization model is one in which the criteria are:

    a. Easily quantifiable
    b. Both qualitative and quantitative
    c. Focused on benefits realization and sustainment
    d. Focused on both short- and long-term goals

Answer Sheet

1.	a	b	c	d
2.	a	b	c	d
3.	a	b	c	d
4.	a	b	c	d
5.	a	b	c	d
6.	a	b	c	d
7.	a	b	c	d
8.	a	b	c	d
9.	a	b	c	d
10.	a	b	c	d

11.	a	b	c	d
12.	a	b	c	d
13.	a	b	c	d
14.	a	b	c	d
15.	a	b	c	d
16.	a	b	c	d
17.	a	b	c	d
18.	a	b	c	d
19.	a	b	c	d
20.	a	b	c	d

Answer Key

1. b.   Ensure compliance with regulatory requirements

   The portfolio manager has a fiduciary responsibility to conform to standards and regulatory requirements.

   Portfolio Management Standard, p. 14

   Task 2 in the ECO in Strategic Alignment

2. b.   Opportunities to be pursued

   The work under way should be validated against organizational strategy updates through strategic alignment analysis. Factors indicating there may be a need to do so include obsolete goals, opportunities to pursue, or responses to regulatory changes.

   Portfolio Management Standard, pp. 44–45

   Task 3 in the ECO in Strategic Alignment

3. b.   Is documented in the portfolio strategic plan

   The organization's performance strategy, along with the communication's strategy and the tolerance for risks, are documented in the strategic plan.

   Portfolio Management Standard, p. 60

   Task 1 in the ECO in Strategic Alignment

4. a.   A simple sum of selected criteria

   This scoring model is easy to construct and use but does not take into account reflective importance among the criteria in the model.

   Milosevic, pp. 22–27

   Portfolio Management Standard, pp. 100–101

   Task 6 in the ECO in Strategic Alignment

5. c.   The components can be quantified

   In a portfolio, all components must be quantifiable so they can be measured, ranked, and prioritized.

   Portfolio Management Standard, p. 3

   Task 6 in the ECO in Strategic Alignment

6. c. Capacity factors such as financial and asset considerations

Capacity is focused on resources, while capability focuses on constraints such as skill set limitations and financial and asset limitations.

Portfolio Management Standard, p. 48

Task 5 in the ECO in Strategic Alignment

7. b. Creating an up-to-date list of components

In order to define the portfolio, an up-to-date list of components is required for regular evaluation, selection, and prioritization.

Portfolio Management Standard, p. 44

Task 4 in the ECO in Strategic Alignment

8. c. Program A

In this example, Program A has the highest number of points with 60 than the other three possible components and should be recommended. Rechenthin, pp. 54–55

Portfolio Management Standard, p. 69

Task 5 in the ECO in Strategic Alignment

9. d. Extent of organizational change

It is necessary to recognize and formalize the organization's ability to handle changes to ensure benefits of portfolio management are realized. The extent of the changes the organization is able to accept then is a key factor used to determine the appropriate mix of components in the portfolio as it relates to strategic objectives, which are inputs to the decision process.

Portfolio Management Standard, p. 27

Task 2 in the ECO in Strategic Alignment

10. a. It is helpful to group stakeholders by category

By grouping stakeholders, one can then best recognize their positions toward the strategic change; determine their levels of power, influence, and interest; and determine the best way to manage their expectations and gain their support.

Portfolio Management Standard, pp. 16, 54, 110

Task 7 in the ECO in Strategic Alignment

11. d. Link project, program, and portfolio principles with organizational enablers

Through these links to the enablers (structural, cultural, technological, and human resource practices), using organizational project management can best support strategic goals.

Portfolio Management Standard, p. 7

Task 7 in the ECO in Strategic Alignment

12. d. Review the portfolio charter

The portfolio charter identifies the portfolio and any sub-portfolios as it considers areas in the organization in scope, hierarchies, and the goals of the portfolio.

Portfolio Management Standard, pp. 64, 66

Task 4 in the ECO in Strategic Alignment

13. a. Review the organization's mission statement

The organization's strategy and objectives are an input to the development of the portfolio strategic plan. The input can be in a document with the mission, vision, and objectives.

Portfolio Management Standard, p. 41

Task 1 in the ECO in Strategic Alignment

14. b. Conduct a readiness assessment

A readiness assessment is a tool and technique in managing strategic change. It determines the if, when, what, and how as to how best to implement the change and move from the 'as is' to the 'to be' state.

Portfolio Management Standard, p. 55

Task 7 in the ECO in Strategic Alignment

15. d. Management approach

The portfolio management plan contains numerous elements. Of significance with these acquisitions, is the need to update this plan because of changes in the management approach; other reasons to update it are changes in priorities and the organizational structure.

Portfolio Management Standard, p. 55

Task 7 in the ECO in Strategic Alignment

16. c.   Portfolio roadmap

The roadmap presents a high-level strategic direction and information in a chronological fashion for portfolio execution.

Portfolio Management Standard, p. 39

Task 8 in the ECO in Strategic Alignment

17. c.   Within the portfolio and between organizational areas

The roadmap shows the internal and external dependencies; the external dependencies are between organizational areas.

Portfolio Management Standard, p. 50

Task 8 in the ECO in Strategic Alignment

18. b.   Contained in the portfolio strategic plan

Among other things, the portfolio strategic plan contains the prioritization model, which is useful as a decision framework to structure components.

Portfolio Management Standard, pp. 46, 48

Task 6 in the ECO in Strategic Alignment

19. c.   Integrate and respond to changes in the portfolio

The portfolio is not static, and organizational strategy changes frequently based on internal and external factors. An adaptive approach to change is required.

Portfolio Management Standard, p. 44

Task 7 in the ECO in Strategic Alignment

20. a.   Easily quantifiable

In a simple model, criteria examples are alignment to strategic goals, expected ROI, investment risks, and dependencies.

Portfolio Management Standard, p. 45

Task 6 in the ECO in Strategic Alignment

Governance

Study Hints

The Governance questions on the PfMP® certification exam constitute 20% of the exam or 34 questions.

These questions are part of the Portfolio Governance Management Knowledge area as well as in the Defining, Aligning, and Monitoring and Controlling Process Groups in *The Standard for Portfolio Management*—Third Edition (2013). There are five processes in Portfolio Governance Management:

1. Develop Portfolio Management Plan
2. Define Portfolio
3. Optimize Portfolio
4. Authorize Portfolio
5. Provide Portfolio Oversight

The first two processes are from the Defining Process Group, Optimize Portfolio is in the Aligning Process Group, and the last two are in the Authorizing and Controlling Process Group.

Effective governance is necessary for portfolio success. Some type of oversight group, such as a Portfolio Review Board, is required for decision making and to ensure the investments analysis is performed to achieve performance targets for the organization.

As the portfolio management plan is developed, it focuses on alignment with the portfolio strategic plan, the charter, and the roadmap. It also draws upon portfolio process assets, organizational process assets, and enterprise environmental factors. As with any plan, it is not a one-time undertaking and is an iterative. This plan also includes as subsidiary plans the performance, risk, procurement, and communications plans. The overall purpose of the plan is to show how the portfolio is defined, organized, optimized, and controlled. Since this plan is critical, it is developed by using a number of tools and techniques. For example, elicitation techniques are used to determine requirements from a variety of sources such as using focus groups and brainstorming sessions, interviews, surveys to involve more stakeholders, and collaboration techniques to poll inputs as a team for consensus, Another useful technique is portfolio organizational structure analysis, as roles and responsibilities for portfolio management need to be determined and assigned.

The portfolio management plan then includes a number of key sections as it explains the governance model to be used, how portfolio oversight will be conducted, describes how the portfolio manager will respond to changes in strategy at the organizational level, and how change control and management will be handled. It also focuses on how the portfolio will be balanced so the mix of components is optimized, and resources are allocated effectively. It further describes any necessary compliance with legislation and the use of the portfolio prioritization model.

Also in the Defining Process Group, an up-to-date list of components in the portfolio is prepared and maintained in the Define Portfolio process. The goal is to ensure resources and investments are focused on those components which have the greatest business value. The portfolio tends to be set up in organization areas and by doing so, common evaluation filters can be developed for evaluation, selection, and prioritization. Qualitative and quantitative information is gathered for each component in the portfolio enabling the organization to balance investments and risks. This process focuses on an evaluation and definition process to promote benefits to be realized by components that are strategically aligned. This process requires a portfolio inventory, which includes descriptive information about each category in the portfolio. Specific evaluation criteria to best support the overall organizational strategy also are determined. Therefore, components are assigned to categories, and weighted ranking and scoring techniques are used. The goal of the scoring models is to evaluate components in a comparable way for effective decision making by the governance oversight group.

Moving to the Aligning Process Group, the Optimize Portfolio process optimizes and balances the portfolio with the goal of performance and value delivery. It evaluates trade-offs of portfolio objectives, managing risks versus return, and balancing short-term and long-term goals. Then, limited resources are balanced across the portfolio reflecting the priorities that are set. Components are balanced with one another in the same category to address different concerns and strategies. Capacity and capability analyses, quantitative and qualitative analyses, weighted ranking and scoring techniques, graphical analysis methods are used. Review Figures 5-11 and 5-12 for examples.

In Monitoring and Controlling, the focus shifts as the portfolio is officially authorized, components are activated, and changes are communicated to affected stakeholders. Resources are officially allocated to these components. A portfolio management information system is used. The Portfolio Oversight process monitors the portfolio to ensure continual alignment and to assess if changes are required for greater business value. The Portfolio Oversight Group and the portfolio manager review information on the components, governance meetings are held, and changes are made based on the results if there are updates to the portfolio.

Following is a list of the major topics in the Governance domain. Use this list to focus your study efforts on the areas that are most likely to appear on the exam.

Major Topics

Portfolio governance management purpose
Defining Process Group
Aligning Process Group
Monitoring and Controlling Process Group
Develop Portfolio Management Plan process

- ■ Purpose
- ■ Inputs
  - – Portfolio strategic plan
    - • Organizational risk tolerance
    - • Communication strategy
    - • Performance strategy
  - – Portfolio charter
    - • Scope, resources, timeline
    - • Stakeholder communications requirements
    - • Performance expectations
    - • Key risks, dependencies, constraints
  - – Portfolio roadmap
    - • High-level strategic direction
    - • Extended timelines
    - • Dependencies
  - – Portfolio process assets
    - • Policies, procedures, knowledge bases
    - • Performance information
    - • Decisions and open issues
    - • Ongoing and planned tasks
  - – Organizational process assets
    - • Policies, procedures, knowledge bases
  - – Enterprise environmental factors
    - • Governance processes
    - • Corporate, economic, governmental variables
- ■ Tools and techniques
  - – Elicitation techniques
    - • Facilitation—focus groups and brainstorming
    - • Surveys—interviews and observations
    - • Collaboration—polls for consensus
  - – Portfolio organization structure analysis
    - • Roles and responsibilities
  - – Integration of portfolio plans
    - • Risk, communications, performance

- Outputs
  - Updates to the portfolio strategic plan and portfolio process assets
  - Portfolio management plan
    - Governance model
    - Portfolio oversight
    - Managing strategic change
    - Change control and management
    - Balancing and managing dependencies
    - Subsidiary plans—performance, communications, risk, and procurement
    - Compliance management
    - Portfolio prioritization model

Define Portfolio

- Purpose
- Inputs
  - Portfolio strategic plan
    - Alignment with strategy
    - Prioritization model
  - Portfolio charter
    - Identifies the portfolio
  - Portfolio
    - Portfolio components
  - Portfolio roadmap
    - Summarizes strategic objectives
    - Identification, categories, ranking and scoring techniques
  - Portfolio management plan
    - Management approach
    - Guidance to evaluate components
  - Portfolio process assets
    - Templates, tools, data, information and guidance
- Tools and techniques
  - Evaluation criteria
  - Portfolio component inventory
    - Descriptors and evaluation criteria
  - Portfolio component categorization techniques
  - Weighted ranking and scoring techniques
    - Single criterion
    - Multiple criteria weighted ranking
    - Multi-criteria scoring model
- Outputs
  - Updates
    - Portfolio
    - Roadmap
    - Portfolio management plan

Optimize Portfolio

- Purpose
- Inputs
  - Portfolio
    - Active components and inactive ones for resource allocation
  - Roadmap
    - 'To-be' state to guide optimization
  - Portfolio management plan
    - Approach to define, optimize, and authorize components
  - Portfolio reports
    - Risks resource pool data, capability and capacity
  - Portfolio process assets
    - Data, tools, templates, information
- Tools and techniques
  - List of components for prioritization
  - Capability and capacity analysis
    - Human, financial and asset
  - Quantitative and qualitative analysis
    - Cost-benefit
    - Spreadsheets—resource loading or cash flow
    - Scenario
    - Probability
    - Strengths, weaknesses, opportunities, and threats (SWOT)
    - Market/competitor
    - Business value
  - Graphical analysis
    - Risk vs. return
    - Pie charts
    - Bubble graphs
- Outputs
  - Updates
    - Portfolio
    - Roadmap
    - Portfolio management plan
    - Portfolio reports
    - Portfolio process assets

Authorize Portfolio

- Purpose
- Inputs
  - Portfolio
    - Previous approved components requiring authorization

- Portfolio management plan
    - Process to authorize components and allocate/reallocate resources
- Portfolio reports
    - Financial, resources, governance decisions
    - Resources—human, shared assets, financial
- Tools and techniques
    - Authorization technique activities
    - Portfolio management information system
        - Show components with assigned resources
- Outputs
    - Updates
        - Portfolio
        - Portfolio management plan
        - Portfolio process assets
    - Reports
        - Resources, assets, governance decisions

Provide Portfolio Oversight

- Purpose
- Inputs
    - Portfolio
        - Current or planned components and organizational strategy
    - Roadmap
        - Integrated view of strategy
    - Portfolio management plan
        - Overall portfolio approach to meet organizational strategy
    - Portfolio reports
        - Performance, resource capability/capacity, risks/issues, financial
    - Portfolio process assets
        - Policies, processes, procedures, knowledge bases
        - Historical performance information, governance decisions, open issues
- Tools and techniques
    - Portfolio review meetings
        - At specified milestones or triggered by external events
    - Elicitation techniques
        - Compiling status reports, facilitating meetings, questionnaires, surveys
- Outputs
    - Updates
        - Portfolio
        - Portfolio management plan
        - Portfolio process assets
        - Portfolio reports

Practice Questions

INSTRUCTIONS: Note the most suitable answer for each multiple-choice question in the appropriate space on the answer sheet.

1. Your organization adopted a management-by projects culture five years ago, and at that time, the CEO set up an Enterprise Project Management Office. Last year, it received the PMO of the Year Award. Now that the organization is moving into portfolio management, the leadership team decided to establish a Portfolio Management Office, and you were selected to manage it. One reason why it felt a Portfolio Management Office should be set up even though the EPMO was working so effectively is that:

    a. It highlights the need for a centralized and structured Governance Board and process
    b. It assists the Governance Board by supporting component proposals and evaluations
    c. In addition to providing guidance on the practice of portfolio management, it also supports component execution
    d. It provides project and program process information to the Governance Board

2. As you work to implement portfolio management in your global financial company, the members of the Portfolio Review Board and other key stakeholders will not be collocated. Meetings will be virtual ones with a variety of social media used for communications regarding the portfolio process, the artifacts, component proposals, reports, and terminations. An example of an enterprise environmental factor as you prepare the portfolio performance plan is:

    a. Organizational culture
    b. Asset capacity
    c. Managing compliance with legislation
    d. Pre-determined risk profile

3. Fortunately, your organization recognized that for portfolio management to be effective, it needed to be managed by someone who was independent of the PMO. It also established a Portfolio Oversight Group with the CEO as its Chair and the heads of the Business Units as its members. While the group has numerous functions, not to be overlooked is making governance decisions in terms of:

   a. Stage gates
   b. Regulatory compliance
   c. Risks and issues
   d. Stakeholder management

4. One of the reasons your CEO is interested in setting up portfolio management in your consulting company is that he came from a background in new product development in which gate reviews were a common practice. However, portfolio management is a new concept in your consulting company as its work is received through competitive bids on proposals or through customer referrals. You also worked in portfolio management with the CEO at the other organization so were hired as the portfolio manager. You are establishing specific decision-making rights and authorities and roles and responsibilities that are needed to manage progress based on:

   a. Benefit realization
   b. Key performance indicators
   c. Critical success factors
   d. Portfolio risk

5. The Chief Financial Officer is a key stakeholder and is a member of your company's Portfolio Review Board. He ensures before each Board meeting that tangible, accurate, and up-to-date financial information is available as portfolio selection and prioritization decisions are made The purpose for his active involvement is to:

   a. Compare project spending with the allocated budget
   b. Ensure the projected financial benefits are quantifiable and delivered
   c. Provide value-for money analyses
   d. Determine if the proposed component will provide sufficient benefits to then acquire external resources to perform it

6. Assume the Portfolio Board met and made a decision to add two programs and three projects to the portfolio and to terminate three other projects as they no longer support the strategic goals. With these newly authorized components, it is time to update the:

   a. Resource availability
   b. Portfolio
   c. Portfolio roadmap
   d. Portfolio strategic plan

7. Your organization has a portfolio performance management plan, and within it benefits realization is described. One of the reasons for doing so is to:

   a. Ensure there is overall portfolio value
   b. Describe key metrics to collect for all aspects of the portfolio
   c. Clearly identify benefits for stakeholders
   d. Have templates in place for benefits realization plans for program and project managers

8. As the portfolio manager it is your responsibility to maintain and update various portfolio process assets. You also are using them as you prepare the portfolio management plan. One that may be overlooked but is helpful in determining portfolio processes is:

   a. Overall governance approach
   b. Portfolio management information system
   c. Existing human resources
   d. Open issues

9. As the portfolio manager you are accustomed to changes and the need to periodically re-prioritize the components as new components are added to the portfolio, others are terminated, and others complete on schedule. As you set up your portfolio performance plan, change management and control is a separate section in it. Within this section you have a:

   a. Process to optimize the component risk
   b. Method to ensure regulatory compliance
   c. Method to schedule change activities
   d. Approach to respond quickly to strategic changes and manage them accordingly

10. As the portfolio manager, you know it is important to communicate to key stakeholders. You need to provide timely messages to those stakeholders who have a component that is approved for authorization into the portfolio and also to the key stakeholders who are involved if a component is removed. You do not want them to learn about these decisions from others, and your focus is on consistent messages. Your approach is documented in the:

    a. Communications strategy
    b. Governance strategy
    c. Portfolio management plan
    d. Portfolio strategic plan

11. An active Portfolio Review Board that meets on a regularly scheduled basis and makes decisions following an agreed upon and defined process so 'pet projects' are not pursued is considered a best practice. After a recent maturity assessment conducted by an outside consultant following *OPM3*, the consultant noted in the Improvement Report that the Portfolio Review Board needed a roadmap for its oversight functions. Having an up-to-date roadmap is helpful as it:

    a. Establishes how the portfolio is defined
    b. Provides information on components for needed changes
    c. Enables actions to be taken to minimize risk
    d. Shows dependencies within the portfolio

12. Assume you are the portfolio manager at the enterprise level in your medical device company. The company's long-standing mission is for its products to be ones of the highest possible quality for its end users. As the enterprise portfolio manager, you set up at the beginning a Portfolio Review Board, and its members are the senior executives of the company. They conduct phase-gate reviews and have a special interest in:

    a. Products and services ranked in the top 10 of all the work under way
    b. Products and services that are ready for commercialization
    c. Research and development initiatives
    d. Quality assurance and quality control

13. In order to keep operating costs low, your executive team is relying on contractors to actually manage about 85% of the work in your organization. You serve as the portfolio manager and often make recommendations, or support those of others, as to whether or not outsourcing is the best approach to follow. The procurement decisions are included in:

    a. The portfolio management plan
    b. The portfolio procurement plan
    c. The portfolio roadmap
    d. The portfolio strategic plan

14. Your government agency, responsible for international espionage, recently had its mandate reduced by the government based on outrage from people throughout the world. As a result, its functions have been reduced, and the agency has greater government oversight. It represents a significant change to the agency, which then impacts the resources and components in the portfolio. The response to such a significant change should be:

    a. In the governance processes
    b. Defined by the portfolio management office
    c. The responsibility of the portfolio manager
    d. In the portfolio strategic plan

15. After a dismal track record of consistently late and over-budget IT programs and projects in your consulting company, the CEO found it was losing proposals given these failures. The CEO realized changes in IT management were required and hired a Chief Information Officer (CIO) with a proven track record of taking companies that were at Level 1 in the Capability Maturity Model Integration (CMMI) to Level 3 in one year. This CIO also recommended that the company implement a portfolio management process to ensure there no unapproved work was under way. The result is that programs and projects now are considered successful, and the company now is winning proposals. However before deciding to bid on a proposal, the company no longer waits for the Request for Proposal (RFP), and now has capture managers present proposals to the Portfolio Review Board when new opportunities are located for approval. These proposals:

    a. Often result in the need for an unscheduled meeting of the Board to be held
    b. Require the same degree of information as that for a program or project
    c. Do not substitute for the need for another proposal if the company wins the contract
    d. Should have the capture manager remain as the sponsor if the contract is awarded

16. Your automotive company is working on its strategic plan for its 2016 model line of vehicles. In the past, when the country's economy was significantly lagging, it had to obtain funds from the government to avoid bankruptcy. Now, the economy has recovered, and the company repaid monies loaned by the government. To ensure it is focusing on the right products and services, the EPMO hired a PfMP to review its portfolio processes and benchmark them against best practices elsewhere. The PfMP consultant noted that the existing portfolio had components that had not been planned effectively. As a result, now the EPMO Director realizes that:

    a. The portfolio prioritization model requires updates
    b. A separate portfolio manager should be hired to oversee the process
    c. Benefits from the existing components are unlikely to be realized
    d. The portfolio governance board needs to be re-constituted into a decision-making forum that meets regularly

17. At your University, as the portfolio manager, you have an inventory, which you strive to maintain, of all the components and have set up key descriptors. Your goal, as you review proposals for new components from sponsors before the final review by the University's Portfolio Committee, is to ensure the proposed component meets requirements for consideration, and this is a first screening. This means, for example, that a component should be one that is:

    a. Able to use existing University resources
    b. Greater than a predetermined minimum size
    c. Focused on quantitative benefits
    d. Independent of current portfolio components

18. After a year of regular Portfolio Review Board meetings at the business unit level, the CEO hired a consultant to determine if the portfolio process was effective or needed some revisions. The consultant recommended that resource capability and capacity information was not up to date as competency profiles were not in place. As a result:

    a. The HR Department now is gathering data on the competencies possessed by staff members
    b. The portfolio management information system needs revision
    c. Portfolio reports require updates
    d. The number of resources and skill set data are to be collected by each business unit

19. Assume you recently were hired at a health insurance company to be its portfolio manager. During the job interview, you learned the company had a portfolio management process in place and a software tool, but the portfolio manager had left the company. You realize your predecessor had not prepared a portfolio management plan, which you believe is critical to success. In preparing it, you want to understand the structure, scope, resources, timeline, stakeholder communication requirements, performance expectations, key risks, dependencies, and constraints so you should:

    a. Conduct interviews with members of the Portfolio Review Board and those responsible for the software tool
    b. Review the portfolio charter
    c. Follow the completed portfolio strategic plan, which was updated two months ago
    d. Review existing portfolio process assets

20. Although portfolio management is a new function in your University, people are embracing it and excited now that a portfolio structure has been set up. The portfolio manager is respected and attained her doctoral degree in project management from the University, and the Chancellor is chairing the Portfolio Review Board, which held its first meeting a week ago and shared the results throughout the University. It now is necessary to:

    a. Update the strategic management plan
    b. Prepare a charter for the Portfolio Review Board
    c. Set up a dashboard system for regular reports from the component managers to the portfolio manager
    d. Authorize the portfolio manager to assign resources to the components

Answer Sheet

1.	a	b	c	d		11.	a	b	c	d
2.	a	b	c	d		12.	a	b	c	d
3.	a	b	c	d		13.	a	b	c	d
4.	a	b	c	d		14.	a	b	c	d
5.	a	b	c	d		15.	a	b	c	d
6.	a	b	c	d		16.	a	b	c	d
7.	a	b	c	d		17.	a	b	c	d
8.	a	b	c	d		18.	a	b	c	d
9.	a	b	c	d		19.	a	b	c	d
10.	a	b	c	d		20.	a	b	c	d

Answer Key

1. a. It highlights the need for a centralized and structured Governance Board and process

   By setting up a Portfolio Management Office, it then highlights why a structured governance process and board are needed especially if none were in existence leading to further benefits, discipline, and understanding of the importance of portfolio management throughout the organization.

   Portfolio Management Standard, p. 18

   Task 1 in the ECO in Governance

2. a. Organizational culture

   Enterprise environmental factors are internal or external conditions that are not under the control of those involved with the portfolio, such as the culture of the organization.

   Portfolio Management Standard, p. 89

   Task 4 in the ECO in Governance

3. c. Risks and issues

   Decisions may be needed concerning risks, both positive and negative, that may affect the components of the portfolio or issues involving existing components especially in terms of resource requirements.

   Portfolio Management Standard, pp. 80–81

   Task 1 in the ECO in Governance

4. d. Portfolio risk

   The risk tolerance of stakeholders involved in portfolio governance is taken into account to consider risk toward the achievement of organizational strategy and governance as part of the governance model.

   Portfolio Management Standard, p. 62

   Task 1 in the ECO in Governance

5. b. Ensure the projected financial benefits are quantifiable and delivered

   Because the portfolio manager considers financial goals and objectives as the portfolio is managed, the finance function is a critical stakeholder and performs up-front component proposal evaluation, monitors the portfolio budgets, compares spending with the allocated budget, and examines realized benefits to ensure financial plan adjustments are made.

   Portfolio Management Standard, p. 13

   Task 3 in the ECO in Governance

6. b. Portfolio

   As an output of the Authorize Portfolio process, the portfolio is updated with the new components, their resources, funding, and any other relevant information.

   Portfolio Management Standard, p. 80

   Task 5 in the ECO in Governance

7. c. Clearly identify benefits for stakeholders

   Stakeholder expectations for portfolio management include the need to understand the benefits of the portfolio, both tangible and intangible, to ensure their continued support for portfolio management.

   Portfolio Management Standard, p. 91

   Task 4 in the ECO in Governance

8. d. Open issues

   Open issues and portfolio management decisions are examples of portfolio process assets in that they are current information to establish portfolio management processes and to define responsibilities.

   Portfolio Management Standard, p. 60

   Task 4 in the ECO in Governance

9. c. Method to schedule change activities

   Change management is handled through a change structure. It involves impact analysis, review and approval or disapproval, prioritization, and scheduling of proposed change actions.

   Portfolio Management Standard, p. 63

   Task 4 in the ECO in Governance

10. c. Portfolio management plan

In addition to describing the process to authorize the portfolio, this plan also describes the key communications needs with a focus on proactive and targeted delivery of messages to key stakeholders.

Portfolio Management Standard, p. 79

Task 4 in the ECO in Governance

11. d. Shows dependencies within the portfolio

The roadmap is useful for Portfolio Oversight as it shows the dependencies between the components. The Portfolio Review Board or similar group then uses it to ensure if changes are made to one component, for example, the changes do not affect other components within the portfolio.

Portfolio Management Standard, pp. 82–83

Task 5 in the ECO in Governance

12. c. Research and development initiatives

Organizations tend to have some form of control for its portfolios. Critical attention is paid in research and development as strategic goals and objectives continually change, and it is then important that the resources involved in research and development are allocated to the most strategic priorities, and that other research and development initiatives that are no longer deemed relevant are terminated.

Portfolio Management Standard, p. 22

Task 5 in the ECO in Governance

13. a. Portfolio management plan

Among other items, this plan includes procurement procedures as it describes the use of procurement to meet the organization's strategic objectives.

Portfolio Management Standard, p. 39

Task 4 in the ECO in Governance

14. a. In the governance processes

The portfolio management plan and the portfolio governance processes outline responses to significant strategic change and how to manage them.

Portfolio Management Standard, p. 63

Task 4 in the ECO in Governance

15. a. Often result in the need for an unscheduled meeting of the Board to be held

   Meetings of the Board tend to be scheduled on a regular basis; however, ad hoc meetings, typically triggered by external events, can be held.

   Portfolio Management Standard, p. 83

   Task 1 in the ECO in Governance

16. c. Benefits from the existing components are unlikely to be realized

   If the organization lacks a successful evaluation and definition process, there may be unnecessary and poorly planned components in the portfolio, hampering benefits realization as components lack strategic alignment.

   Portfolio Management Standard, p. 64

   Task 3 in the ECO in Governance

17. b. Greater than a predetermined minimum size

   Many organizations have limits as to which components must be officially approved to be part of the portfolio. For example, a small project that takes less than 40 hours to complete with one person doing the work typically would be excluded.

   Portfolio Management Standard, p. 67

   Task 2 in the ECO in Governance

18. c. Portfolio reports require updates

   Resource capacity and capability reports are examples of the type of reports that may need to be updated as a result of the Provide Portfolio Oversight process.

   Portfolio Management Standard, p. 84

   Task 2 in the ECO in Governance

19. b. Review the portfolio charter

   These items should be part of the portfolio charter, and it is required to prepare the portfolio management plan.

   Portfolio Management Standard, p. 60

   Task 4 in the ECO in Governance

20. a. Update the strategic management plan

    The structure identifies the portfolio, sub-portfolios, programs, and projects based on the organization areas included, hierarchies, timelines, and goals. As they are part of the strategic plan, once the structure is defined, the strategic plan requires updates.

    Portfolio Management Standard, p. 47

    Task 1 in the ECO in Governance

Portfolio Performance Management

Study Hints

Effective portfolio performance is essential to portfolio management success. The Portfolio Performance questions on the PfMP® certification exam constitute 25% of the exam or 43 questions.

These questions are part of the Portfolio Performance Management Knowledge area as well as in the Defining and Aligning Process Groups in *The Standard for Portfolio Management*—Third Edition (2013). There are three processes in Portfolio Performance Management:

1. Develop Portfolio Performance Management Plan
2. Manage Supply and Demand
3. Manage Portfolio Value

The first process is from the Defining Process Group, while the other two are in the Aligning Process Group.

The purpose of portfolio performance management is to ensure there is an optimal mix of components in the portfolio, and they are sequenced to best achieve business value. Organizational value then is obtained through this process of planning, measuring, and monitoring overall performance and sourcing needed resources. Performance metrics, therefore, are used to determine areas of improvement, which include both quantitative and qualitative measures. In many organizations the tangible metrics will take precedence, while in others intangible metrics may be of greater value.

As the portfolio performance management plan is developed, it focuses on how portfolio value is defined and how resources are allocated to the components. This plan may be part of the portfolio management plan, or it may be a separate document. To develop it, the goals from the portfolio strategic plan are used along with objectives to reach these goals Through performance management progress toward these goals are measured, and any needed changes to the portfolio mix to reach these goals are assessed, The benefits of each component also are emphasized and tracked for the optimum value Another area of emphasis is providing the governance group and other key stakeholders with metrics to determine if the organizational strategy will be met. As well, the metrics provide sponsors and component managers the details on how their components are doing in the context of others in the portfolio as the goal is to ensure the portfolio consists of the components that will create the greatest value given available resources.

Selection of appropriate metrics is based on the areas of importance to each organization, and a goal is to have a few key metrics that can be easily analyzed and used. They are ones that follow the SMART (specific, measurable, attainable realistic and time based) approach. It therefore is useful to have an easy-to-use portfolio management information system and to focus on capacity and capability analysis

As the plan is prepared, a number of people are involved, and templates and examples are included for performance measures and targets. These key performance indicators (KPIs) then report on the progress of the portfolio as well as future indicators.

In the Aligning Process group, managing supply and demand is critical. Review Figure 6-4 in *The Standard* and recognize the usefulness of a master schedule of resource allocation. Supply, as used in *The Standard*, involves resource capacity, recognizing resources are more than just people, while Demand is the resource requirements from proposed and existing components in the portfolio. It is essential to minimize unused capacity and demand, and balance resources according to priorities since resources are constrained in most organizations. Optimal monitoring of the supply and demand then is critical to portfolio success. Scenario analysis, quantitative and qualitative analysis, and capability and capacity analysis are used, and through this process updates to the portfolio are expected.

This knowledge area also involves Managing Portfolio Value, and it is expressed in different ways by organizations. A best practice is to use a value measurement framework to organize how value will be created and measured, considering both tangible and intangible metrics. The expected value from each component in the portfolio is measured throughout its life cycle to see if changes are warranted to achieve greater benefits. The objective is to ensure through the current work is in the portfolio, objectives are being achieved, and value estimates are established, focusing on continuous improvement of the value measurement framework, Recommendations for change to the portfolio management mix are expected through this process. Study Figure 6-9 as an example of a performance variance report that may be used, along with Figure 6-10 for a benefits realization plan. Changes in weighting for scoring performance also may be necessary, as shown in Figures 6-11 and 6-12. Value scoring and measurement analysis and benefits realization analysis are used. Not to be overlooked is the usefulness of the portfolio efficient frontier, as shown in Figure 6-13 as a tool to optimize the portfolio given resource constraints. Consider reviewing the work done by Henry Markowitz in the financial arena and relate it to portfolio management for the organization.

Following is a list of the major topics in the Portfolio Performance domain. Use this list to focus your study efforts on the areas that are most likely to appear on the exam.

Major Topics

Portfolio performance management purpose

- Optimal mix of components
- Sourcing of key resources
- Examples of quantitative and qualitative measures

Defining Process Group
Aligning Process Group
Develop Portfolio Performance Management Plan process

- Purpose
- Inputs
  - Portfolio management plan
    - Stakeholder expectations and requirements
    - Governance model
    - Framework for strategic change
    - Planning, procurement, and oversight processes
  - Portfolio process assets
    - Portfolio strategic plan
    - Charter
    - Schedules
    - Processes, policies, procedures, knowledge bases
  - Organizational process assets
    - Human-resource related policies
  - Enterprise environmental factors
    - Organizational culture
    - Structure
    - Human resources
- Tools and techniques
  - Establishing SMART performance metrics
  - Portfolio management information system
    - Cost, performance metrics, risks
    - Performance improvement measures
    - Established baselines
  - Capability and capacity analysis
    - Definitions, resource types
    - Resource schedules
    - What-if scenarios
    - Finite capacity planning and reporting
    - Resource management tools

■ Outputs
  – Portfolio management plan updates
    • Steps and measures to manage portfolio performance
    • Roles and responsibilities for measurement
    • Performance measures
    • Performance reporting
    • Resource optimization
    • Benefit realization
    • Key performance indicators
  – Portfolio process assets updates

Manage Supply and Demand

■ Purpose
  – Definitions
  – Unused capacity and unmet demand
  – Types of resources
  – Continual monitoring
■ Inputs
  – Portfolio
    • Resource requirements
    • Component prioritization
  – Portfolio management plan
    • High-level guidelines for management
    • Reporting risks, communicating, engaging stakeholders, recommending changes
  – Portfolio reports
    • Resource utilization
    • Vacation schedules
    • Equipment/supply
    • Financial
■ Tools and techniques
  – Scenario analysis
    • Possible resource allocation
    • Impact on component schedules
    • Resource management tools
  – Quantitative and qualitative analysis
    • Resource leveling
    • Resource schedules
    • Dependency analysis
    • Trend analysis
  – Capability and capacity analysis
■ Outputs
  – Updates to the portfolio, portfolio management plan, portfolio reports

Manage Portfolio Value

■ Purpose
  - Methods to express value
  - Value measurement framework
  - Change in expected value of components
  - Recommending portfolio changes
■ Inputs
  - Roadmap
    • Component delivery high-level timeline
  - Portfolio management plan
    • Identification and assessment of value
    • Monitoring benefit interdependencies
  - Portfolio reports
    • Aggregate component performance
    • Variance reports
    • Alert techniques
    • Standardized formats
    • Burn-down/burn-up reports
    • Traffic light reports
■ Tools and Techniques
  - Enabler for optimization and authorization
  - Benefit criteria examples
  - Elicitation techniques
    • Stakeholder meetings
    • SWOT analysis for benefits
  - Value scoring and measurement analysis
    • Scoring models—benefits
    • Revision of scoring weights
    • Cost/benefit analysis for benefits
    • Comparative advantage analysis
    • Progress measurement techniques—earned value
    • Value measurement techniques
    • Portfolio efficient frontier—Henry Markowitz, best possible level of return for the level of risk
  - Benefit realization analysis
    • Results chain—cause-and-effect relationships
    • Outcome probability analysis of the portfolio and success criteria, cumulative distribution
■ Outputs
  - Updates
    • Portfolio management plan
    • Portfolio reports
    • Portfolio process assets

Practice Questions

INSTRUCTIONS: Note the most suitable answer for each multiple-choice question in the appropriate space on the answer sheet.

1. All portfolio decisions, including approvals, prioritizations, and artifacts such as business cases, should be maintained. A best practice to follow is to:

   a. Include them in the organization's knowledge management system
   b. Set up a portal dedicated to portfolio management
   c. Include them as part of the portfolio management information system
   d. Set up a document version control system

2. Entering a new market requires careful analysis to determine whether or not it is profitable to do so. Assume you are proposing to sponsor a new product line for your company. You are recommending it to be part of the portfolio. You have completed a market and competitor analysis. To ensure the capacity is available, you decide to also:

   a. Determine resource loading requirements
   b. Assess available asset capacity
   c. Use the Delphi approach to solicit views of experts
   d. Prepare a histogram

3. With resources in short supply, especially with people with specialized skill sets or the need for certain types of equipment, your resource management software system ideally should have the capability to:

   a. Use resource leveling
   b. Use resource smoothing
   c. Enable sharing of resource calendars
   d. Show soft versus hard booking

4. Developing metrics for portfolio management takes time to ensure that the portfolio manager has a preapproved set of metrics that include quantitative and qualitative information. As well, the portfolio manager must aggregate these metrics from the components in order that the stakeholders are not overwhelmed by the metrics and can use them for informed decision making This set of metrics should monitor a number of items including:

   a. Benefits
   b. Value proposition
   c. Customer relationship management
   d. Risk profile

5. There are a number of features that are common to portfolio components regardless if one is working in a public sector or private sector organization. You are managing the portfolio at your University and are explaining how it is set up to a new member of the Board of Trustees. You explain that:

    a. You have grouped the components in the portfolio so they can be managed more effectively
    b. The portfolio has been set up in a way that resources are allocated to each project or program when it is part of the portfolio
    c. The functional departments in the University basically are only involved if they are pursuing new projects or programs such as redesigning how the admissions process is handled
    d. Each component in the portfolio is based on a current investment by the University

6. As the portfolio manager in your company, which is now entering a new product line to use robots to deliver packages, you have been in this role for two years. Progress has been slow at times, but the executive team now has embraced it. Detailed proposals must be prepared for new components, which are then ranked and scored, and portfolio balancing occurs quarterly. During this rebalancing, if components are terminated because others have a higher priority, and resources are scarce, those involved seem surprised. This shows:

    a. Training guides in portfolio management would be helpful
    b. Resource competency development requires more attention
    c. Different portfolio reports should be submitted by component managers
    d. Component managers should attend rebalancing sessions

7. Assume you work for a recording company. You are to recommend one new component to be optimized in the company's portfolio. You only can select one because of resource constraints. You can select Project A to develop a web site, Program A to enter the music video market, Program B for a major nationwide tour, or Project B for new T-shirts. You have net present value (NPV) data available to assist you in making your recommendation as follows:

Project A NPV	*Program A NPV*	*Program B NPV*	*Project B NPV*
5% = 5,243	5% = 2,320	5% = 6,800	5% = 3,000
10% = 2,841	10% = 1,254	10% = 3,275	10% = 2,755
15% = 1,563	15% = 688	15% = 1,679	15% = 700

You recommend:

a. Project A
b. Program A
c. Program B
d. Project B

8. It is important to ensure the key stakeholders are supportive of strategic portfolio changes, and stakeholder identification is not a one-time activity especially in your transit authority in which a new CEO tends to be appointed every two years. As the portfolio manager, when you identify the key stakeholders who should be involved or at least consulted, one objective in doing so is to determine their:

a. Roles and responsibilities
b. Power and urgency
c. New pain points
d. Reactions

9. Assume in your Portfolio Management Office, you are using a mathematical model to forecast future outcomes using historical results. Such an approach has proven to be especially useful in terms of predicting whether components in the portfolio will be completed on time and/or within budget especially for companies in which time to market is critical and do not want schedules to slip. You decided to apply it to resource allocation as it can provide information:

a. On required resources
b. To show the effect on resource capacity
c. To show needed resource competencies not available in the company
d. To see if resource adjustments are warranted

10. It is important not only to select the portfolio that has the greatest value strategically but also to implement it. It becomes especially difficult when financial resources are limited and are a constraint after performing capacity analysis. One approach to consider is:

   a. NPV
   b. ROI
   c. Comparative advantage analysis
   d. Resource allocation

11. In most organizations there are more proposed components to consider, but there are constraints as to how much new work can be undertaken. Fortunately, as the portfolio manager, you maintain an up-to-date list of components in progress and receive status reports that you use when it is time to meet with your Portfolio Review Board. At the upcoming meeting, 15 possible components are being proposed, but only five can be selected. This is based on:

   a. Value delivery
   b. Mandatory requirements
   c. Financial analysis
   d. Benefits identification

12. While your solar panel and heating company has had steady business over the last three years, the company wants to pursue additional markets as well as maintain its existing market share. You are the portfolio manager for this $50 million dollar company. However, financial resources are scarce for new initiatives, and you are putting in place a prioritization methodology. Your goal is to understand the capabilities of the various software available rather than letting the tool 'become the fool'. You decided one approach would be to focus on problem identification, such as the cost of capital being high, and present a series of issues to understand systemic explanations, which means you are focusing on:

   a. Analytical hierarchy process
   b. Prioritization algorithms
   c. Constraint management
   d. Simulation techniques

13. Assume you were asked by the director of your business unit to evaluate how best to allocate scarce resources to existing components as he wants to know if the resources are over allocated or have time available to support new components in the portfolio. The best approach in terms of capability and capacity analysis is to:

    a. Evaluate existing competencies by using a competency analysis and perform a gap analysis
    b. Use a SWOT analysis in terms of existing resources
    c. Use finite capacity planning and reporting
    d. Evaluate resource availability by using the PMIS

14. Establishing categories is a best practice in portfolio management especially when making prioritization decisions and in resource allocation. Assume your company, which produces a variety of ice machines to assist people quickly recover from shoulder, ankle, and knee surgeries, set up eight different categories for its components with one being for mandatory requirements. When the Portfolio Review Board met on Friday, it made a decision to add two components from the new products category and to terminate three components in the research and development category. When you reviewed these decisions, you found there were dependencies that had not been considered, which also included:

    a. Cost/benefit dependencies for the entire group
    b. Resource implications given the specialized resources in the R&D category
    c. A short-term trend rather than taking a longer-term view of possible future benefits and their sustainability
    d. The need to immediately call another meeting of the Review Board to escalate your concerns

15. Your organization conducted an *OPM3®* assessment of its portfolio management practices and wanted to do so to evaluate the current portfolio management structure to see if it was adequate and had the right allocation of resources. The *OPM3®* Certified Professional found that among other things there was limited emphasis on portfolio management except in one business unit. People in the one business unit had training on why portfolio management was needed and had a defined process in place. After reviewing the assessor's report, it seemed obvious that:

    a. This one business unit's leadership team should be commended for their work
    b. This one unit's processes should be mandated across the company
    c. The level of management commitment is uneven
    d. More attention is needed on portfolio resource availability against the integrated schedule in the company

16. Assume you were hired as the portfolio manager for the Inspector General's Office of your government agency. While this Office must do all inquiries according to its mandate, portfolio management still is useful as in the portfolio performance management process, as it then can:

    a. Make resource sourcing decisions
    b. Identify opportunities and threats
    c. Achieve performance targets
    d. Assess changes and dependencies

17. With the impact of new technology and the continual need to do more with less, change in strategy is common in organizations. What worked well in the past may not be suited to the dynamic environment. As strategy shifts this means:

    a. A gap is typical between the 'as is' state and the 'to be state'
    b. The original portfolio charter serves as a guideline
    c. Interdependency analysis is required
    d. Capability and capacity analysis should be performed

18. Your low acid canned food company has come under scrutiny by regulatory authorities lately as some of its well-known products have had to be recalled as they contained botulism. Your executives decided the company needed to change its name to avoid customer dissatisfaction and pursue other markets. You have been appointed as the portfolio manager, a new function for this company. Since this is a new function to define your role, a best practice is to review the:

    a. Portfolio roadmap
    b. Portfolio strategic plan
    c. Portfolio charter
    d. Portfolio management plan

19. Recently, the CEO in your company retired, and the Board of Directors appointed a new management team. Because of competition from overseas companies, also specializing in farm equipment such as tractors and crawlers, the new management's challenge is to be less risk adverse than in the past and to focus on new markets and opportunities. You are remaining as the portfolio manager, and the new management team is the Portfolio Review Board. As they meet to focus on these new components, you know they will be interested in:

    a. Competitor intelligence
    b. Measures to maximize portfolio value
    c. The current portfolio categories
    d. The method used to rate and score proposals

20. Working to manage the portfolio value, you plan to provide a list in priority order of those components you feel should be considered. You have decided to use benefit realization analysis as one method to do so as in your pharmaceutical company some products may never be commercialized. Fortunately, you have a definition of value that is aligned with strategy. You may want to estimate potential portfolio outcomes with respect to success criteria. You will express your results by:

    a. Gaps and overlaps to be addressed
    b. A cumulative distribution
    c. Earned value
    d. ROI and NPV

Answer Sheet

	a	b	c	d
1.	a	b	c	d
2.	a	b	c	d
3.	a	b	c	d
4.	a	b	c	d
5.	a	b	c	d
6.	a	b	c	d
7.	a	b	c	d
8.	a	b	c	d
9.	a	b	c	d
10.	a	b	c	d

	a	b	c	d
11.	a	b	c	d
12.	a	b	c	d
13.	a	b	c	d
14.	a	b	c	d
15.	a	b	c	d
16.	a	b	c	d
17.	a	b	c	d
18.	a	b	c	d
19.	a	b	c	d
20.	a	b	c	d

Answer Key

1. c.   Include them as part of the portfolio management information system

The portfolio management information system contains tools and techniques to gather, integrate, and disseminate outputs of portfolio processes. Among other items that can be part of this system are tools and techniques involving documents—as a repository and as a version control system.

Portfolio Management Standard, pp. 25–26

Task 10 in the ECO in Performance

2. d.   Prepare a histogram

Resource schedules a part of capability and capacity analsyis. They are histograms that combine and detail forecasts and ongoing resource supply and demand..

Portfolio Management Standard, p. 90

Task 8 in the ECO in Performance

3. d.   Show soft versus hard booking

Soft booking indicates resources are authorized to components based on the date another component is expected to end, while hard booking indicates they are committed to the component once it is authorized to be part of the portfolio. It helps maximize resource use.

Portfolio Management Standard, p. 93

Task 7 in the ECO in Performance

4. d.   Risk profile

Risk profiles of the decision makers, such as in the Portfolio Oversight Group, can change. Metrics can be set up to monitor risk profiles based on those at the component level.

Portfolio Management Standard, p. 89

Task 6 in the ECO in Performance

5. a. You have grouped the components in the portfolio so they can be managed more effectively

   The various components of the portfolio have common features. One of them is to permit the organization to group them for effective management.

   Portfolio Management Standard, pp. 4–5

   Task 6 in the ECO in Performance

6. a. Training guides in portfolio management would be helpful

   This situation is an example of a lesson learned, an output of the Manage Portfolio Value process, as it shows some in the organization lack a detailed understanding of the overall portfolio management process.

   Portfolio Management Standard, p. 104

   Task 10 in the ECO in Performance

7. c. Program B

   Using NPV as a way to recommend components for optimization, a dollar one year from today is worth less than a dollar today. The more the future is discounted, or the higher discount rate, the less the NPV of the component. If the NPV is high, the component then is ranked high, leading to Program B.

   Milosevic, pp. 42–44

   Portfolio Management Standard, pp. 74, 102

   Task 7 in the ECO in Performance

8. c. New pain points

   As the portfolio undergoes strategic changes, areas of interest to key stakeholders also may change in the process. Therefore, one purpose is to see if there are any new pain points that require attention.

   Portfolio Management Standard, p. 54

   Task 5 in the ECO in Performance

9. d. To see if resource adjustments are warranted

   Trend analysis is useful as a tool and technique in the Manage Supply and Demand process. It can show if resource requirements consistently have been underestimated or if resources are consistently over or under performing.

   Portfolio Management Standard, p. 95

   Task 7 in the ECO in Performance

10. c. Comparative advantage analysis

    Comparative advantage analysis is an example of a value scoring and measurement analysis technique used in the Manage Portfolio Value process. It has value as competing efforts may reside internal or external to the portfolio. It may also include a "what-if" analysis.

    Portfolio Management Standard, p. 102

    Task 7 in the ECO in Performance

11. c. Financial analysis

    A number of different capacity and capability analyses can be done, one of which involves financial resources to ensure funding is available for components to be selected.

    Portfolio Management Standard, pp. 74, 102

    Task 7 in the ECO in Performance

12. c. Constraint management

    Constraint management is an approach in which the focus is on identifying the root causes that deter companies from making investments of financial resources, time, and effort. It provides an approach to ease constraints and encourage the organization from making decisions to stimulate growth.

    Portfolio Management Standard, p. 15

    Task 6 in the ECO in Performance

13. c. Use finite capacity planning and reporting

    This approach is useful since it indicates any resource bottlenecks and over and under resource allocations.

    Portfolio Management Standard, p. 90

    Task 7 in the ECO in Performance

14. a. Cost/benefit dependencies for the entire group

Optimization decisions must consider dependencies including those that involve cost and benefit dependencies for the entire portfolio. A variety of analyses are recommended before these decisions are made.

Portfolio Management Standard, p. 102

Task 6 in the ECO in Performance

15. c. The level of management commitment is uneven

This scenario shows that portfolio management lacks the needed depth and breadth of commitment to portfolio management by the board of directors, executives, and senior executives.

Portfolio Management Standard, p. 23

Task 1 in the ECO in Performance

16. a. Make resource sourcing decisions

While a number of activities are done in the portfolio management process, a major one is to source key resources, finance, assets, and human resources, for optimal returns.

Portfolio Management Standard, p. 85

Task 7 in the ECO in Performance

17. a. A gap is typical between the 'as is' state and the 'to be' state

When strategy changes, any gap should be analyzed to determine whether realignment in resources or adjustment in the components in the portfolio are required.

Portfolio Management Standard, p. 88

Task 8 in the ECO in Performance

18. d. Portfolio management plan

Among other things, the portfolio management plan provides a framework for strategic change.

19. b. Measures to maximize portfolio value

Measures are needed to ensure optimal resource performance and maximum portfolio value. They should be relevant to overall strategy and objectives.

Portfolio Management Standard, p. 89

Task 9 in the ECO in Performance

20. b. A cumulative distribution

Benefit realization analysis is a tool and technique in the Manage Portfolio Value process. This is an example of using outcome probability analysis of the portfolio. A cumulative distribution can be used to set realistic targets aligned with stakeholder risk tolerances.

Portfolio Management Standard, p. 103

Task 6 in the ECO in Performance

Portfolio Risk Management

Study Hints

Similar to risk management at the project or program levels, it is even more important at the portfolio level. Many of the same tools and techniques can be used, but at a higher level, and others are more sophisticated since the portfolio risks are numerous and are from many different sources. The Portfolio Risk Management questions on the PfMP® certification exam constitute 15% of the exam or 25 questions.

These questions are part of the Portfolio Risk Management Knowledge area as well as in the Defining and Aligning Process Groups in *The Standard for Portfolio Management*—Third Edition (2013). There are two processes in Portfolio Risk Management:

1. Develop Portfolio Risk Management Plan
2. Manage Portfolio Risks

The first process is from the Defining Process Group, and Manage Portfolio Risks is in the Aligning Process Group.

It should be noted that the definition of portfolio risk in *The Standard* on page 119 is slightly incorrect, and instead the definition in the Glossary section should be followed. The purpose of portfolio risk is to assess and manage any risks to the portfolio and strive to focus on exploiting opportunities and minimize threats that may affect the portfolio. It also emphasizes interdependencies between risks between components, especially between the high-priority ones. Through risk management, new components may be identified to consider to be added to the portfolio.

At the portfolio level, risk management focuses on short-, medium- and long-term risks. It differs from that at the program or project level with the long-term view as the organization may decide to embrace a risk as it may view it as one with a high reward. Such an approach is especially noted in organizations working on complex new products in which the desired technology may not be available as expected. Therefore, at the portfolio level the focus is to increase the possibility of positive events while decreasing any that may adversely affect the portfolio considering value, the strategic fit, and portfolio balance.

This approach means active consideration of the organization environment to evaluate its management practices and approaches, the number of concurrent components, and the dependency on stakeholders external to the organization for component and overall portfolio success. It then is critical in portfolio risk management to determine the root cause of any negative risks and correct it as well as to capitalize on potential positive opportunities.

Reserves are needed at the portfolio level because of possible threats, and their management is a responsibility of the portfolio manager. The portfolio manager also can in some cases aggregate risk responses through common characteristics and often relies on equity protection. The emphasis is on risk planning, risk assessment, and response.

As well, external, structural, and execution risks may arise and require consideration. Anyone can identify risks, and the culture should be one in which open communication about risks is encouraged. But, people at different levels also will have different perspectives regarding risk and different concerns to address. Issues of transparency, organizational integrity, and corruption are other considerations especially since risks at the portfolio level are greater than the sum of those at the program or project levels. Risk thresholds and overall organizational attitudes toward risk require identification to see if the organization is risk adverse, tolerant or risk taking. Since internal portfolio risks are greater if they occur, resource commitments then may be required for positive or negative risks.

The portfolio risk management plan describes the approach that will be followed with references to any applicable organizational policies and is used to assess new risks in proposed components as well as overall portfolio risk. The plan also shows how the portfolio oversight group will balance investments against expected return for known risks to support risk-based decisions.

In developing the plan, it is useful to review the portfolio management plan, portfolio and organizational process assets, and enterprise environmental factors. Useful tools and techniques include weighted ranking and scoring techniques. These are helpful especially in organizations with multiple portfolios and can be applied to technical and management risks. High-level risk plans can be defined by these techniques along with cost elements and schedule activities and needed risk contingency reserves. Graphical analytical techniques also are useful such as a probability/impact matrix as shown in Figure 8-5 in *The Standard*. Other considerations include importance, timing, interdependencies, confidence limits, and prioritized risk lists. A number of different quantitative and qualitative techniques can be used such as status and trend analysis, rebalancing methods, different investment approaches, and risk exposure charts. These charts, for example, can provide useful information such as the outcome probability analysis of the portfolio and the probability of achieving the portfolio's objectives, as shown in Figure 8-6.

The risk management plan then describes the methodology that will be followed, specific roles and responsibilities, risk measures, how often the risk management process will be performed, and categories, with examples as shown in Figure 8-7.

As the plan is completed, the Manage Portfolio Risks process is under way continuously. It consists of risk identification, analysis, responses, and overall monitoring and control. This process involves review of the overall portfolio and development of a portfolio risk register, similar to that at the program or project level as well as an issue register. It also involves reviewing the portfolio management plan, reports, portfolio and organizational process assets, and enterprise environmental factors.

As in the planning process, weighted ranking and scoring techniques may be useful to evaluate any existing risks and identify any new ones. Quantitative and qualitative analyses also are helpful. One approach is to prepare a Tornado Diagram, see an example in Figure 8-10, to show each identified risk's contribution or impact to strategic objectives. Other approaches are net present value, sensitivity analysis Monte Carlo simulations, investment choice analysis, and variance analyses.

This process often leads to a need to repeat the entire risk management processes and modify existing plans. Portfolio reports in addition to the risk register are helpful, such as the example shown in Figure 8-11.

Following is a list of the major topics in the Portfolio Risk Management domain. Use this list to focus your study efforts on the areas that are most likely to appear on the exam.

Major Topics

Portfolio risk management purpose

- Assess and analyze portfolio risks
- Capitalize on opportunities
- Minimize threats
- Assess interdependencies
- Determine risk versus reward
- Provide overall risk reserves
- Consider
  - External risks
  - Internal risks
  - Structural risks
  - Execution risks
- Identify risks by people throughout the organization
- Determine the organization's risk threshold
- Assess resource needs

Defining Process Group
Aligning Process Group
Develop Portfolio Risk Management Plan process

- Purpose
- Inputs
- Portfolio management plan updates
  - Governance model
  - Performance management
  - Communication
  - Stakeholder engagement
- Portfolio process assets
  - Components
  - Selection criteria
  - Prioritization algorithms
  - Risk register
  - Issue register
  - Performance matrices
  - Resources
  - Budget
- Organizational process assets
  - Vision and mission
  - Strategy and objectives
  - Risk tolerance
  - Lessons learned

- Enterprise environmental factors
  - Organization culture and structure
  - Organizational project management
  - Legal and regulatory
  - Industry requirements
  - Market conditions
  - Published information
- Tools and techniques
  - Weighted ranking and scoring
    - Important portfolio risks, owners, tolerance, process
  - Graphical analytical methods
    - Probability and impact matrix
    - Importance, timing, interdependencies, confidence limits, prioritized lists
  - Quantitative and qualitative analysis
    - Status and trend analysis
    - Rebalancing methods
    - Investment choice tools
    - Risk exposure charts
- Outputs
  - Portfolio management plan updates
  - Portfolio risk management plan
    - Methodology
    - Roles and responsibilities
    - Measures
    - Frequency
    - Categories
  - Portfolio and organizational process asset updates

Manage Portfolio Risks

- Purpose
  - Identify risks
  - Analyze risks
  - Develop risk responses
  - Monitor and control risks
- Inputs
  - Portfolio
    - Risk register
    - Issue register
  - Portfolio management plan
    - Roadmap, funding, technical knowledge, guidelines, budget
  - Portfolio reports
    - Performance
    - Governance decisions

- Status
- Trends
- Capacity and resource use
- Funding/budget
- Strategic alignment
  - Portfolio process assets
    - Process assets from stakeholders
    - Knowledge bases, lessons learned, historical information
  - Organizational process assets
    - Vision and mission
    - Strategy, objectives, values
  - Enterprise environmental factors
    - Commercial data bases, academic studies, benchmarking, industry studies
- Tools and techniques
  - Weighted ranking and scoring techniques
  - Quantitative and qualitative analysis
    - Interdependencies, importance, timing, confidence limits
    - Quantitative analysis
      - NPV, estimated NPV, ROI, Payback, IRR
    - Sensitivity analysis
      - Tornado diagram
    - Modeling and simulation
      - Monte Carlo
    - Qualitative analysis
      - Probability and impact
      - Assumptions analysis
      - Influence diagrams
      - Risk-portfolio component chart
      - Heat maps
    - Investment choice analysis
      - Changes in strategic goals
      - Gaps in investments in the portfolio
    - Variance analysis
    - Performance review measurements
- Outputs
  - Updates
    - Portfolio management plan
    - Reports
    - Portfolio process and organizational assets

Practice Questions

INSTRUCTIONS: Note the most suitable answer for each multiple-choice question in the appropriate space on the answer sheet.

1. In preparing a portfolio risk management plan, obviously risk management is embedded as part of portfolio management. Planning, though, is helpful as it identifies areas around risk management, one of which is:

    a. Prioritization algorithms
    b. Risk register
    c. Performance metrics
    d. Risk owners

2. Assume you are working to manage and control risks that affect the portfolio. As you do so, you need to review the portfolio and its risks and issues. An issue is:

    a. A risk that should occur
    b. A condition that leads to a risk
    c. Defined by root causes
    d. An event with an impact on the portfolio

3. A number of different tools and techniques from the Manage Portfolio Risks process have been useful to gather data for the Portfolio Governance Board in order for its members to then:

    a. Realign the budget to better manage risks
    b. Assess the effectiveness of the risk responses
    c. Determine whether sufficient risk planning has been useful
    d. Consider other methods of identifying threats and turning them into opportunities

4. Recently, your organization was acquired by a larger competitor. While your organization will retain its name, this acquisition is a major risk to the existing portfolio. It is of such a magnitude that:

    a. The portfolio risk management process should be repeated
    b. Communications to different stakeholder groups should be issued
    c. The next step is a risk assessment for the organization
    d. It is doubtful that many of the current components in the portfolio will be continued

5. Assume you are the portfolio manager for your bank, which has offices in about half of the states in your country. You are focusing on those items in which effective risk planning and risk responses lead to more positive outcomes for the bank. One way in which you can show a standard view and assessment of organizational risks is to:

    a. Use a probability and impact matrix
    b. Assess trends in terms of portfolio value
    c. Determine the probability of achieving the objectives of the portfolio
    d. Determine the combination of the most effective investment choices

6. Change is constant whether one is working on an operational activity, on projects or programs, or in a portfolio management position. Risks often result because of unplanned changes that may cause the entire portfolio to change. As a portfolio manager, it is important to best identify any changes that could occur, and one can do so by:

    a. Discussing potential changes each time the Portfolio Review Board meets
    b. Setting up a process as part of the portfolio management plan for an integrated approach across the organization's various portfolios to best manage changes
    c. Establish and maintain a change register
    d. Monitor internal and external environment changes

7. The risk owner is listed for each risk in the risk register. Together with the portfolio manager, risk owners can:

    a. Determine a fallback plan
    b. Choose the most effective response strategy
    c. Develop contingency plans and trigger conditions
    d. Select a mix of risk response strategies

8. You are considering a possible investment choice tool to use as you work to develop the portfolio risk management plan. Assume in your organization, the effects of portfolio velocity is critical, which means you should use:

    a. Sensitivity analysis
    b. Market variability analysis
    c. Rebalancing
    d. Time-to-market variability

9. In order to keep up with and hopefully surpass the competition in your health insurance company, you realize your portfolio needs some new components. There have been a large number of customer complaints about the difficulty of signing up for supplemental insurance, the long waits on the phone, the poor and unfriendly web site, and the difficulty in refilling prescriptions. Your company's executive team has decided it has to invest in new technology. However, rather than only purchase a slight upgrade, the executives have elected to use new but unproven sophisticated technology that will ensure upgrades then will not be needed for ten years. This situation shows:

   a. The willingness to take risks based on high rewards
   b. The need to ensure that the selected vendor has a definite commitment to provide service to sustain the promised benefits
   c. An emphasis on the long term as it is a low risk but high reward approach
   d. Once the technology is purchased, its installation requires significant work in portfolio strategy

10. Your grocery store is considered to be a mid-range store in that it is large, is not a discount store, and also is not focused exclusively on high-end organic products. It selects products that consumers desire and discontinues others that people seem not to purchase. However, it has stores throughout 15 different states, and prices and products offered change constantly. If the store decides to offer a new product or a new service, such as a health care center, or discontinue a service, such as the delicatessen, the risks of doing so must be considered carefully as part of investment analysis. As the portfolio manager, you therefore focus on:

    a. Benchmarking studies
    b. Time-to-market
    c. Market-payoff variability
    d. Market requirement variability analysis

11. The Portfolio Review Board maintains a record of its decisions. This means that if it increases the portfolio budget to cover preventive actions to the portfolio, such as acquiring additional resources to enhance time to market, once this action is taken, it is necessary to update the:

    a. Roadmap
    b. Portfolio strategic plan
    c. Portfolio management plan
    d. Portfolio performance plan

12. Your construction company has experienced numerous delays during the winter since the majority of its work is in the northern US and the UK. The severe winter has meant many projects have been canceled or delayed. Your management team, who are the members of the Portfolio Review Board, plan a special meeting next Friday to specifically address weather-related risks recognizing the investment threats and then the possible lack of resources once the projects can proceed. During this meeting to evaluate the existing risks and to add any new ones, a useful tool and technique is:

    a. Investment choices
    b. The efficient frontier
    c. Weighted ranking and scoring
    d. Probability and impact assessments

13. Your chain of restaurants has built its reputation on outstanding customer service. It led the use of hand held credit card readers to speed the time to receive a check and pay rather than looking at a check, providing a credit card, giving it to a server, receiving the card and the check to sign, and then obtaining a receipt. Now it is implementing bitcoins as a way to hasten the checkout process. To get customers to use bitcoins, it is offering a 50% discount on all food and drinks that are offered. It is providing mobile apps to customers to use them. This is an example of a(n):

    a. External positive risk
    b. Internal negative risk
    c. Overall portfolio risk
    d. Risk tolerant company

14. There are a number of criteria to consider in identify portfolio stakeholders. One criterion that often is overlooked is:

    a. Regulatory requirements
    b. The need to involve consumer groups
    c. The importance of sustainability
    d. Risk management strategies

15. Assume you have worked with your cross-functional team of stakeholders to identify the key risks affecting the asphalt company's portfolio. There were five negative risks: inclement weather, health and safety concerns, lack of resources, lack of funding, and inability to complete programs and projects on time. A positive risk was increased demand for the product. You are maintaining a portfolio risk register. A next step is to:

    a. Assign a risk owner
    b. Review the negative risks for root causes and assumptions
    c. Determine potential responses for each identified risk
    d. Set up triggers for the negative risks

16. Assume you have prepared your portfolio risk management plan and have determined responses to risks to the portfolio that might occur whether they are threats or opportunities. For those risks considered to be the most critical to the portfolio, such as a competitor releasing a similar product earlier than planned by your company, decreasing employee morale shown by many people leaving to take positions elsewhere resulting in lack of needed resources, or a subcontractor defaulting that is supporting several prime contractors, one way to display this information to key stakeholders is by:

    a. Maintaining a complete risk register and highlighting the critical risks
    b. Using a cause-and-effect diagram
    c. Preparing a sensitivity analysis
    d. Using Monte Carlo

17. Assume you are planning to present the results of your portfolio risk analysis at the next meeting of the Portfolio Review Board. You plan to show a comparison of the relative importance and impact of variables with a high degree of uncertainty and to those that are more stable. A useful approach is to prepare:

    a. A probability/impact matrix
    b. An expected monetary value analysis
    c. A tornado diagram
    d. An assumptions analysis

18. As you work to manage portfolio risks, you decide to review the portfolio management plan. It is helpful in identifying risks as it:

    a. Provides a list of key governance decisions
    b. Lists the root causes of previously identified risks
    c. Lists key stakeholders that may influence the risk posture
    d. Requires an understanding of the portfolio roadmap

19. Assume you are the CEO of a small, boutique portfolio, but you are risk adverse, and your profile is dominated by one high risk component as shown in the following graph. Your goal is to move to a target portfolio, which can be characterized as medium risk even though its expected return is lower. This graphic shows:

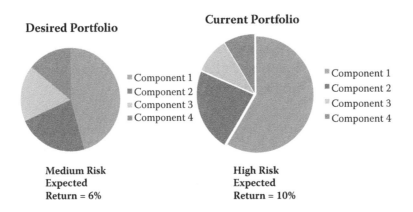

Desired Portfolio — Medium Risk Expected Return = 6%

Current Portfolio — High Risk Expected Return = 10%

- a. The importance of balancing risks versus rewards
- b. The problem of having only a few components in the portfolio
- c. The need for greater quantitative analysis approaches to calculate overall return
- d. The need to focus on investment choices

20. Assume you are one of five owners in a start-up company, and your plans are to enter the low calorie food market, given the number of obese people of all ages; provide easy-to-heat and nutritious food to military personnel stationed overseas; enter the ice cream market with new and delicious flavors; and also offer dark chocolate candy. You have put together your strategic plan and have hired people to lead the new product offerings plus people for internal operations. Each new product development head has at least 20 components. This is a situation in which:

- a. Resource allocation is critical
- b. The quantity of components requires balancing
- c. The quality of portfolio management may be a structural risk
- d. Market leadership is a key to portfolio success

Answer Sheet

1.	a	b	c	d
2.	a	b	c	d
3.	a	b	c	d
4.	a	b	c	d
5.	a	b	c	d
6.	a	b	c	d
7.	a	b	c	d
8.	a	b	c	d
9.	a	b	c	d
10.	a	b	c	d

11.	a	b	c	d
12.	a	b	c	d
13.	a	b	c	d
14.	a	b	c	d
15.	a	b	c	d
16.	a	b	c	d
17.	a	b	c	d
18.	a	b	c	d
19.	a	b	c	d
20.	a	b	c	d

Answer Key

1. d.   Risk owners

Other areas around risk management that are due to planning include portfolio risks affecting the organization, risk tolerance, and risk processes. They are part of weighted rankings and scoring techniques, a tool and technique in the Develop Risk Management Plan process.

Portfolio Management Standard, p. 125

Task 2 in the ECO in Risk Management

2. d.   An event with an impact on the portfolio

Issues differ from risks and are different at the portfolio level than at the program or project levels. It is an event with a current impact on the portfolio. Negative risks if they are not identified and treated become issues; and positive risks if not identified or treated are lost opportunities.

Portfolio Management Standard, p. 131

Task 4 in the ECO in Risk Management

3. a.   Realign the budget to better manage risks

The purpose of collecting and reviewing the portfolio risk information is to enable the governance group to evaluate the portfolio to determine whether rebalancing components, resources, or the budget is required to realign portfolio risks.

Portfolio Management Standard, p. 132

Task 6 in the ECO in Risk Management

4. c.   The next step is a risk assessment for the organization

This situation is a need to update organizational process assets, an output to the Manage Portfolio Risks process.

Portfolio Management Standard, p. 135

Task 6 in the ECO in Risk Management

5. a. Use a probability and impact matrix

   Risk probability or likelihood assessments are used to show the possible occurrence of a specific risk, while risk impact or consequences assessments can show the potential effect of positive risks or opportunities and negative risks or threats. They can be displayed graphically in a probability and impact matrix, a tool and technique in the Develop Portfolio Risk Management Plan process.

   Portfolio Management Standard, p. 126

   Task 2 in the ECO in Risk Management

6. d. Monitor internal and external environment changes

   The portfolio manager has a broader role in change management than a project or program manager as he or she continuously monitors changes in the environment, both internal and external.

   Portfolio Management Standard, p. 6

   Task 1 in the ECO in Risk Management

7. d. Select a mix of risk response strategies

   The portfolio manager along with the risk owners can work together to determine effective response strategies, while the risk owner on his or her own focuses on the specific assigned risk.

   Portfolio Management Standard, p. 134

   Task 5 in the ECO in Risk Management

8. d. Time-to-market variability

   The purpose of this tool is to determine the effects of portfolio velocity, which means there is overall corporate pressure to adopt technology challenges as organizations strive for breakout growth and innovation in the face of increased complexity and an increase in the rate of change.

   Portfolio Management Standard, p. 127

   Task 2 in the ECO in Risk Management

9. a. The willingness to take risks based on high rewards

   One way that risk management differs at the portfolio level than at the program or project levels is the organization may elect to enhance risks for high rewards. In this situation, the technology may not work, but if it does work, its acquisition will lead to enhanced market share and greater profits.

   Portfolio Management Standard, p. 120

   Task 1 in the ECO in Risk Management

10. d. Market requirement variability analysis

    This investment choice is one in which it analyzes changes in market requirements in relation to the entire portfolio as shown in this situation, as an investment choice tool used in the Develop Portfolio Risk Management Plan process.

    Portfolio Management Standard, p. 127

    Task 2 in the ECO in Risk Management

11. c. Portfolio management plan

    This update is an output of the Manage Portfolio Risks process to ensure this plan is updated as the agreed-upon actions are updated and monitored.

    Portfolio Management Standard, p. 134

    Task 5 in the ECO in Risk Management

12. c. Weighted ranking and scoring

    Weighted ranking and scoring techniques are a tool and technique used in the Manage Portfolio Risks process. As part of the Authorizing and Controlling Process, the Portfolio Review Board or comparable group, may use these approaches during a regular meeting or a special meeting to concentrate solely on risks.

    Portfolio Management Standard, p. 132

    Task 4 in the ECO in Risk Management

13. d. Risk tolerant company

    Some organizations are risk adverse, while others are risk tolerant. This company is moving quickly into a new way of working as a way to provide greater benefits to customers and promote market leadership.

    Portfolio Management Standard, p. 123

    Task 1 in the ECO in Risk Management

14. d. Risk management strategies

    The size of the organization and program and project management practices are considerations in identifying portfolio stakeholders, but certain stakeholders are identified based on the portfolio's goals and its risk management strategies.

    Portfolio Management Standard, p. 26

    Task 1 in the ECO in Risk Management

15. b. Review the negative risks for root causes and assumptions

    Once a negative risk is identified, it should be reviewed using root cause analysis to determine if there are other associated risks and also to see if the risk can be avoided once the cause is identified. As well, assumptions are a source of risk and require review using assumption analysis techniques.

    Portfolio Management Standard, pp. 130–131

    Task 4 in the ECO in Risk Management

16. a. Maintaining a complete risk register and highlighting the critical risks

    The risk register is used throughout portfolio management. A column can be added to show if it is a critical risk, and reports for stakeholders can focus on the status of these critical risks, the date of possible and actual impact, recovery actions, and the date the risk is closed.

    Portfolio Management Standard, pp. 37, 131

    Task 4 in the ECO in Risk Management

17. c. A tornado diagram

    The results of a sensitivity analysis can be displayed in a tornado diagram to show the parameters that lead to a high degree of variability and those with a lesser effect.

    Portfolio Management Standard, pp. 132–133

    Task 5 in the ECO in Risk Management

18. d.   Requires an understanding of the portfolio roadmap

The portfolio roadmap shows the various components and their inter-dependencies, among other things, which can be sources of risks. As well, the portfolio management plan also requires an understanding of overall portfolio funding, including the budget allocated to risk management, and technical knowledge regarding the portfolio components.

Portfolio Management Standard, p. 130

Task 4 in the ECO in Risk Management

19. d.   The need to focus on investment choices

Investment choice analysis is a tool and technique in the Manage Portfolio Risks process as investment choices show portfolio alignment and can show any gaps in terms of investment to the portfolio as these gaps may be risks.

Portfolio Management Standard, p. 133

Task 5 in the ECO in Risk Management

20. c.   The quality of portfolio management may be a structural risk

Structural risks involve the organization's ability to organize its portfolio. This situation is an example of over-ambitious plans in diverse areas, and inconsistent strategies required to realize them, which may be threats to portfolio success.

Portfolio Management Standard, p. 122

Task 3 in the ECO in Risk Management

Communications Management

Study Hints

For portfolio management to be effective in organizations at any level, communications management is essential so people realize why it is necessary. The Communications Management questions on the PfMP® certification exam constitute 15% of the exam or 25 questions.

These questions are part of the Portfolio Communications Management Knowledge area as well as in the Defining and Aligning Process Groups in *The Standard for Portfolio Management*—Third Edition (2013). There are two processes in Portfolio Communications Management:

1. Develop Portfolio Communications Management Plan
2. Manage Portfolio Communications

The first process is from the Defining Process Group, and Manage Portfolio Information is in the Aligning Process Group. These processes transcend the other knowledge areas in portfolio management since communications emphasize the importance of ensuring stakeholder expectations for information about portfolio management in the organization are met. Consistent and transparent communications lead to effective portfolio decisions. Communications also are a key competency for the portfolio manager as he or she must be able to communicate effectively with stakeholders at any level, both internal and external. The importance of communications dictates the requirement to prepare and follow a communications strategy.

Portfolio communications involve a two-way dialogue between stakeholders such as executives, functional managers, sponsors, program managers, project managers, suppliers and other vendors, regulatory agencies and other interest and consumer groups. As the communications management plan is prepared, stakeholder identification and analysis are needed for it to be useful. Even though stakeholders have been identified once a decision is made to embrace portfolio management, identification is an ongoing process. As stakeholders are identified, then their information needs are compiled and maintained, recognizing these information needs are much broader than those at a program or project level.

In developing the communications management plan, it is helpful to review the proposed and ongoing portfolio components to consolidate and standardize communications at the portfolio level. The roadmap is helpful since it shows, among other things, interdependencies between components. The portfolio management plan may include the communications plan as a subsidiary document but also is useful as if it changes, the changes may lead to new stakeholders with new information requirements. Portfolio performance reports show the total investment on each component to indicate its value, and dashboards often are used. Portfolio process assets cannot be overlooked.

Tools and techniques focus on stakeholder analysis to best recognize those stakeholders with the greatest interest in or influence in the portfolio investment. Review the types of stakeholders as shown in Table 7-1 in *The Standard*, and their levels of influence and interest as displayed in Figure 7-5, in which stakeholders are assigned into groups. It then follows that a stakeholder matrix to list roles, interests and expectations by the type of stakeholder, as shown in Table 7-2, should be prepared so no one is overlooked.

Since the portfolio manager then spends much of his or her time interacting with these stakeholders, he or she can determine whether their interests have changed, and in meeting with them, can get feedback on the communications plan. These data are useful to then develop a communications matrix, such as the one in Table 7-3, to array the different communications methods to use, their frequency, the intended recipients, and the communications vehicles.

The communications plan then defines the process to gather the needed information and distribute it and to set expectations for effective communications; the risks of lack of communications cannot be underestimated.

However, once the plan is prepared, the information must be managed, which involves collecting, analyzing, storing, and distributing it to stakeholders. Often, web portals are used for ease of access, and information must be up to date and of high quality for it to have value to the various stakeholders. In managing information, the portfolio with the list of components, the portfolio management plan to ensure communications are aligned with the other processes, and portfolio reports are reviewed.

Tools and techniques focus on data gathering in component review meetings and operational status reports and the portfolio management information system with a document repository and document version control system to best capture and manage portfolio information. Communications requirements analysis continues especially in market comparative analysis and to ensure meaningful forecasts can be prepared. Different communications methods will be used such as dashboards, as shown in Figure 7-8, to provide extensive information in a short format, resource histograms, and communications calendars. As portfolio information is managed, updates may be required to the portfolio management plan, the roadmap, and the portfolio process assets.

Following is a list of the major topics in the Communications Management domain. Use this list to focus your study efforts on the areas that are most likely to appear on the exam.

Major Topics

Portfolio communications management purpose

- Satisfy important information needs of stakeholders
- Promote informed portfolio decision making
- Minimize risks of insufficient communications

Defining Process Group
Aligning Process Group
Develop Portfolio Communications Management Plan process

- Purpose
- Inputs
  - Portfolio
    - Component information to consolidate and standardize communications
  - Roadmap
    - Interdependencies may impact communication objectives
  - Portfolio management plan
    - Initial stakeholder list
    - Shows communication requirements in all elements of the plan
    - Changes introduce new stakeholders
  - Portfolio reports
    - Information on total component investments
    - Use of dashboards
  - Portfolio process assets
    - Portfolio manager roles and responsibilities
    - Status reports
    - Risk profiles and assessments
    - Forecasts and variances
    - Governance, funding, resource decisions
    - Portfolio value assessments
    - Delegation of responsibilities for communications
- Tools and techniques
  - Stakeholder analysis
    - Interest and influence in portfolio investments
    - Information about each stakeholder and relationships
    - Types of stakeholders
    - Assignment into groups
    - Stakeholder matrix
  - Portfolio manager's relationships and interactions with stakeholders
  - Communications requirements analysis
    - Reviews for redundant information
    - Communications matrix
    - Organizational culture

■ Outputs
  – Portfolio management plan updates
    • Communications management plan
      ■ Objectives
      ■ Roles and responsibilities
      ■ Stakeholders and expectations
      ■ Methods to collect and store information
      ■ Methods to access and deliver information
      ■ Frequency
      ■ Policies and constraints
  – Portfolio process assets updates

Manage Portfolio Information

■ Purpose
■ Inputs
  – Portfolio
    • New components and new stakeholders
  – Portfolio management plan
    • Align communications with other processes
  – Portfolio reports
    • Assess stakeholder requirements
  – Component reports
    • Impact on overall portfolio performance
  – Portfolio process assets
    • Requirements, technology, media, record retention policies, security
    • Communications guidelines and procedures, distribution methods, risks, performance data
■ Tools and techniques
  – Elicitation
    • Collect data and information to distribute
  – Portfolio management information system
    • Document repository, version control, technology based, capture and manage portfolio communications
  – Communications requirements analysis
    • Market comparative analysis, data value, prepare forecasts
  – Communications methods
    • Dashboards, resource histograms, communications calendars
■ Outputs
  – Updates
    • Portfolio management plan
    • Reports
    • Portfolio process assets

Practice Questions

INSTRUCTIONS: Note the most suitable answer for each multiple-choice question in the appropriate space on the answer sheet.

1. Working for a government regulatory agency, there is a focus on government in the "sunshine" meaning regulatory meetings are open to anyone who wants to attend. As the portfolio manager in this area, you want the communications process to be a transparent one so consistent messages regarding portfolio management are provided. As well, transparent communications promote:

    a. Benefits realization
    b. An emphasis on sustainability
    c. Optimal resource allocation
    d. Effective prioritization

2. During your stakeholder analysis, you put stakeholders into groups and first determined if they were internal or external. You then focused on determining their level of interest and influence concerning the portfolio. You asked questions such as:

    a. What are the needed benefits?
    b. What is their stance on change?
    c. Do they want to be informed of all developments?
    d. Are they concerned about updates on key risks?

3. Assume you have been reviewing the various reports you provide to your stakeholders and realized some were not meeting their needs now that portfolio management has been implemented in the organization. You have an automated PMIS and can use it to quickly consolidate data from components. You met with some key stakeholders and updated three reports and realized two others were no longer of interest. Now you need to:

    a. Update your PMIS
    b. Update component reports
    c. Update the portfolio management plan
    d. Update the portfolio performance plan

4. Your executives recently attended a conference and learned that your company is lacking because its portfolio management practices and resources are inadequate. The executives learned that research reports showed an effective portfolio management program leads to completing programs and projects on time and under budget and increases ROI. However, in setting up a portfolio process, a key item is:

   a. A benefits realization plan
   b. A communications model
   c. Information on progress and results
   d. Managing scope

5. You set up a stakeholder matrix to show influence and interest. Recently, the director of manufacturing and also the director of human resources have shown an interest in the portfolio, the Portfolio Oversight Group, how often decisions are made, and how prioritization is done. You classified them in your matrix as people to inform on an occasional basis but ones with a high level of possible interest. This example shows:

   a. It is necessary to provide more information consistently to those in this category
   b. The portfolio manager requires continual engagement with stakeholders
   c. They should receive information on all portfolio decisions
   d. They should be considered for openings on the Portfolio Oversight Group if an existing member leaves

6. As the portfolio manager, most of your time is spent in communications with stakeholders. You are concerned you may not be meeting their needs as they seem to be requesting a number of ad hoc reports. You decided to meet with representatives from each stakeholder group to assess lessons learned for communications effectiveness. You next should:

   a. Use the results from these sessions to revise the communication management plan
   b. Update portfolio process assets
   c. Consider more effective ways to provide stakeholders with needed information
   d. Revisit roles and responsibilities for communications to increase timeliness in responding to requests

7. Working as a portfolio manager, you realize you are spending even more time communicating than when you were a program and project manager especially since portfolio management is a new function in your state government agency. In order to realize the full value of each component in the portfolio, you need to:

   a. Ensure it is representative of the investments in it made by the Steering Committee
   b. Has the needed resources to be able to realize its business benefits
   c. Manage interfaces with operations
   d. Ensure each project in the portfolio is set up to create deliverables to support organizational objectives

8. In your liquefied natural gas company, portfolio management has been implemented for eight years. As a result, the company has found it to be beneficial in many ways especially in terms of investment decisions and resource allocation. The company has a large number of portfolio process assets based on its eight years of work in this area, and one used in communications is:

   a. Security requirements
   b. Knowledge management
   c. Open issues
   d. Vision and mission statements

9. Assume you are a staff member in the recently established Portfolio Management Office in your non-profit organization. The Portfolio Manager reports to the CEO. Your role is to prepare a communications strategy. To do so, you surveyed stakeholders at all levels to meet important information needs. An objective of this strategy is to:

   a. Meet requirements set forth in the portfolio charter
   b. Provide information that leads to effective decisions
   c. Set the stage for a detailed communications plan
   d. Ensure people in the organization are committed to portfolio management

10. Assume you are the portfolio manager in your pharmaceutical company, and you also are the director of the Portfolio Management Office. You report to the CEO of the company. Your responsibilities are varied, and one that may be overlooked is:

    a. Lessons learned
    b. Financial standing
    c. Benefits and outcomes
    d. Risks and issues

11. Assume you are in the process of determining how to best collect, analyze, deliver, and store information about the portfolio to stakeholders. You want to set up the most effective approach to best manage portfolio information. As you do so, you want to make sure all communication action plans are aligned with the:

    a. Strategic goals and objectives
    b. Portfolio management plan
    c. Portfolio strategic plan
    d. Portfolio performance plan

12. Your company hired an *OPM3®* Certified Professional to conduct an assessment of its portfolio management processes against the best practices in the *OPM3®* model. The consultant found only 10 of the possible portfolio management best practices were being followed and put together a prioritized improvement plan to focus on a sequence of initiatives to follow to achieve desired outcomes and other best practices. It was evident by the few best practices that were in place was that there is a/an:

    a. Need to address misinformation regarding portfolio management and its purpose
    b. Requirement to review the existing portfolio reporting processes and procedures
    c. Over emphasis on program and project management versus portfolio management in the company
    d. Requirement to assess the resources devoted to portfolio management

13. A goal in portfolio communications and in managing information is to promote informed decision making. There are many ways to ensure that data to be provided to stakeholders are meaningful. One approach is to:

    a. Establish a PMIS
    b. Determine communications constraints
    c. Conduct lessons learned sessions
    d. Select data to share on a web portal

14. Increasingly, many organizations are using portals or more sophisticated internal knowledge management systems as a way to maintain information and make it easy to access. You are, however, working in an avionics company in which the founder, who is 93, still comes to work, and many of the executive team members have been with the company 40 to 50 years. It did, however, diversify its portfolio and is working now as a software provider for a major aviation company as well as continuing its successful product line. The executives meet monthly to review the portfolio and make resource allocation and investment decisions. You were hired as a portfolio manager and are striving to implement new processes and procedures. Since it is a new function, you are involving others and preparing a portfolio communication plan. In this situation, you should:

    a. Realize stakeholders prefer reports
    b. Set up an internal portal to store information
    c. Use dashboards distributed via e-mail
    d. Suggest that the avionics company move into knowledge management as its next step

15. Assume you are the portfolio manager. You have completed your stakeholder analysis and have classified them into groups to best manage their communications expectations and influence as well as to recognize their interest in the portfolio. You have a small staff of four reporting to you, and you decided to assign one of your team members to work with each stakeholder group. One team member met with you today and was concerned that:

    a. The group lacks interest in the portfolio
    b. Some stakeholders have different interests
    c. Some stakeholders seem to require constant attention, and other assigned work is suffering
    d. The stakeholder group desires more of your attention, not that of the team member

16. Assume you have prepared your communications strategy for portfolio-related information, and your strategy emphasizes transparency in terms of portfolio priorities and status. This approach provides:

    a. Support for the Portfolio Oversight Group
    b. Credibility for the portfolio manager
    c. A basis for a portfolio management information system
    d. An approach to manage stakeholders with varying degrees of interest

17. After finishing your stakeholder identification and analysis of their expectations, you next decided to prepare a communications requirements analysis in order to:

    a. Determine communications frequency, recipients, and vehicles
    b. Evaluate communication policies, assumptions, and constraints
    c. Determine roles and responsibilities to manage communications
    d. Establish communication objectives

18. As the portfolio manager in your dairy cooperative, you must communicate effectively with stakeholders both internal and external to the company and to your members. Your goal is to provide stakeholders with the information they require, and you must understand the components in the portfolio to:

    a. Evaluate communications at the portfolio level
    b. Simplify information distribution
    c. Ensure both external and internal stakeholders receive comparable information
    d. Enable access by any stakeholder to information about portfolio components

19. Assume that your organization has set up a number of portfolio reports, and the component managers provide them to the portfolio manager. While the organization has a number of policies and plans for portfolio management, it lacks a formal communication management plan. You are now preparing this plan, and you want to make sure you have information to show the total investment in each component because:

    a. It communicates the assessed value of the portfolio
    b. It provides a schedule as to when funding is required for each component
    c. If a decision is made to terminate a component, the sunk costs are known
    d. It shows the total cost of ownership of the portfolio

20. Your goal as a portfolio manager is to develop a strong communications management plan to keep interested stakeholders informed about your progress in portfolio management. Although you have reached out to numerous stakeholders, you know other portfolio processes also can help in this process such as:

    a. Strategy
    b. Finance
    c. Governance
    d. Risk

Answer Sheet

1.	a	b	c	d
2.	a	b	c	d
3.	a	b	c	d
4.	a	b	c	d
5.	a	b	c	d
6.	a	b	c	d
7.	a	b	c	d
8.	a	b	c	d
9.	a	b	c	d
10.	a	b	c	d

11.	a	b	c	d
12.	a	b	c	d
13.	a	b	c	d
14.	a	b	c	d
15.	a	b	c	d
16.	a	b	c	d
17.	a	b	c	d
18.	a	b	c	d
19.	a	b	c	d
20.	a	b	c	d

Answer Key

1. c. Optimal resource allocation

   Given that resources are scarce, through transparent communication, there is consistency in how resources are allocated to the components in the portfolio to better ensure the needed resources are available according to the component's priority.

   Portfolio Management Standard, p. 107

   Task 2 in the ECO in Communications

2. b. What is their stance on change?

   This question is useful in assessing influence to determine if the stakeholder resists change, support change initiatives, or is considered a thought leader or early adopter.

   Portfolio Management Standard, p. 110

   Task 3 in the ECO in Communications

3. c. Update your portfolio management plan

   Over time, reporting requirements will change. The portfolio communication plan is a subsidiary plan to the portfolio management plan, and updates to the portfolio management plan are an output of the Manage Portfolio Information process.

   Portfolio Management Standard, p. 118

   Task 5 in the ECO in Communications

4. c. Information on progress and results

   The portfolio manager recognizes how the portfolio relates to the organization's strategy and plays a key role in implementing the strategy by monitoring the initiation of initiatives in the plan and communicating progress and results.

   Portfolio Management Standard, p. 15

   Task 3 in the ECO in Communications Management

5. b. The portfolio manager requires continual engagement with stakeholders

   Communications requirements will change over time, and some stakeholders may want more information and involvement, while others may feel they do not need as much information as they are receiving. The portfolio manager uses engagement to ensure information needs are met.

   Portfolio Management Standard, p. 112

   Task 3 in the ECO in Communications

6. b. Update portfolio process assets

   These lessons learned sessions facilitate knowledge gained on overall communication management and then lead to the need to update portfolio process assets, an output of the Develop Portfolio Communication Management Plan process.

   Portfolio Management Standard, p. 113

   Task 6 in the ECO in Communications

7. c. Manage interfaces with operations

   Since processes and deliverables used by operations management may be outputs of the components of the portfolio, the portfolio manager manages relationships and interfaces with operations to ensure the full value of the portfolio is realized.

   Portfolio Management Standard, p. 5

   Task 3 in the ECO in Communications Management

8. a. Security requirements

   Knowledge about security requirements is useful in the Manage Portfolio Information process as portfolio process assets are an input to this process.

   Portfolio Management Standard, p. 116

   Task 5 in the ECO in Communications

9. b. Provide information that leads to effective decisions

   It is easy to collect information, but to support collecting it people need to see that it is analyzed and used.

   Portfolio Management Standard, p. 105

   Task 2 in the ECO in Communications

10. a. Lessons learned

Portfolio management is continually evolving, and there is a need to focus on continuous improvement in its performance. Lessons learned are a key area of interest for the Portfolio Management Office and should be collected and analyzed frequently.

Portfolio Management Standard, p. 111

Task 3 in the ECO in Communications

11. b. Portfolio management plan

It is an input to the Manage Portfolio Information process since it provides the intended approach for portfolio management.

Portfolio Management Standard, p. 115

Task 5 in the ECO in Communications

12. a. Need to address misinformation regarding portfolio management and its purpose

The assessment is complete and shows limited best practices in portfolio management. It leads to a need for better communications management regarding why portfolio management is important, and it may require work to best meet stakeholder expectations.

Portfolio Management Standard, p. 23

Task 1 in the ECO in Communications

13. d. Collect data to share on a web portal

One purpose in using communications requirements analysis as a tool and technique for the Manage Portfolio Information process is to select meaningful data for communicating in newsletters or in news areas on a web portal.

Portfolio Management Standard, p. 117

Task 5 in the ECO in Communications

14. a. Realize stakeholders prefer reports

The organization's culture is significant in determining communication formats and frequency. In this situation, since this company is slowly changing, and is not that comfortable with technology, reports are a more useful option to convey portfolio information.

Portfolio Management Standard, p. 113

Task 4 in the ECO in Communications

15. b. Some stakeholders have different interests

   Stakeholder identification and analysis are ongoing. Even though stakeholder groups are established, over time, the interests of the stakeholders in the group may change.

   Portfolio Management Standard, p. 110

   Task 3 in the ECO in Communications

16. b. Credibility for the portfolio manager

   A transparent approach shows the portfolio manager respects stakeholder views at all levels, not solely those of the governance group, as he or she wants to involve people at all levels for portfolio success.

   Portfolio Management Standard, p. 105

   Task 2 in the ECO in Communications

17. a. Determine communication frequency, recipients, and vehicles

   This analysis focuses on how best to communicate information to stakeholders and how often they need it. It also helps to make sure the most effective communication vehicle is used.

   Portfolio Management Standard, p. 112

   Task 4 in the ECO in Communications

18. a. Evaluate communications at the portfolio level

   Communications at the portfolio level requires different information for stakeholders than that provided at the component level. The portfolio is a key input to the Develop Portfolio Communications Plan process.

   Portfolio Management Standard, p. 108

   Task 2 in the ECO in Communications

19. a. It communicates the assessed value of the portfolio

   Portfolio reports are an input to the Plan Communication Management process. This type of performance report is essential to communicate the value of the portfolio for effective governance decisions.

   Portfolio Management Standard, p. 109

   Task 2 in the ECO in Communications

20. d. Risk

Drawing from other portfolio management processes such as performance and risk can help in the development of the communications management plan.

Portfolio Management Standard, p. 107

Task 2 in the ECO in Communications

Practice Test 1

This practice test is designed to simulate PMI®'s 170-question PfMP® certification exam. You have four hours to answer all questions.

INSTRUCTIONS: Note the most suitable answer for each multiple-choice question in the appropriate space on the answer sheet.

1. Ideally, the organization practices a policy of open communications on risks and encourages people to point any out at all levels, even if the risk does not affect one's own work and especially if it affects the portfolio. Different people, though, have different perspectives of the various portfolio risks based on their position in the organization. Assume a risk has been identified concerning the organization's operating model. This risk was identified by:

   a. A functional manager
   b. A PMO Director
   c. An executive
   d. The portfolio manager

2. In your portfolio some of the programs and projects that are being pursued will realize benefits throughout the program and project's life cycle, while others will not realize the benefits until the program or project is closed or years later. This means as the portfolio manager, you should:

   a. Prepare a portfolio benefit realization plan
   b. Set up KPIs to document progress in benefit realization
   c. Include portfolio benefits, results, and expected value in the portfolio strategic plan
   d. Distribute regular reports on benefit realization as part of the portfolio communications management strategy

3. Your organization, given the economic downturn in your country, decided to reduce its staff by 90% and outsource all operational activities including those of program and project managers and their teams. It has, however, retained the portfolio manager, and it has a Portfolio Review Board comprised of senior executives that meets monthly. Since outsourcing is the norm and not the exception, the manager of Procurement and Contracting is a major stakeholder. Her areas of interest are:

   a. Benefits and outcomes toward strategic goals
   b. Overall portfolio performance
   c. Financial standing
   d. Change decisions

4. Your company has a stated policy that all stakeholders are to be treated in an ethical manner. It is one of the largest project management training firms in the world and is a Registered Educational Provider with the Project Management Institute as well as with other associations. It is active in portfolio management to ensure it is offering the most beneficial products and services, both leading edge and traditional, to its customers. Its policy toward its stakeholders is:

   a. Documented in the stakeholder expectations plan
   b. Considered as a legitimate right
   c. Part of the portfolio communications strategy
   d. A portfolio governance process

5. Assume your airline just merged with a competitor, making it the largest in the country. Your company has a defined portfolio management process in place, which is considered effective by your executives. The other airline uses a more informal approach. Executives from both airlines now have different philosophies concerning new work to be pursued to be the leading airline in the country. In terms of the Authorize Portfolio process, this means:

   a. The complete portfolio process will require changes
   b. The portfolio requirements should be evaluated
   c. The portfolio management information systems will require consolidation
   d. An outsider should be hired as the portfolio manager for objectivity

6. Assume your automotive company is new to formal portfolio management. It has had for years a strategic plan and tries to be first to market for new and improved features on its vehicles each model year. You were hired as the portfolio manager to provide a more disciplined approach for determining new products to pursue as well as existing ones that should be terminated. So far, you have set up an approach, established categories for the various components, and determined a method to rank and score new proposals for consideration. Now you are working to set up practices to follow to optimize the portfolio. In doing so, it is important to note that:

    a. The criteria to optimize the portfolio may be the same as that used in the scoring model

    b. A portfolio management information system should be set up

    c. Future investment requirements are a key criterion to consider

    d. Compliance with organizational standards cannot be overlooked

7. Each time a strategic change occurs, it requires a number of updates, and it includes the need to update the portfolio process assets including:

    a. Timelines

    b. Prioritization model

    c. Lessons learned

    d. Communication requirements

8. Assume you are helping the Portfolio Review Board select and implement the portfolio with the best alignment to strategy as you work to create a list for to be considered for prioritization. One possible component on the surface does not seem to be one that is profitable, but you believe that over time changes will occur that will make it cost/beneficial to pursue. This means you are using:

    a. Market analysis

    b. Business value analysis

    c. Scenario analysis

    d. Options analysis

9. Recognizing that different components can have different types of risks, you decide to see how each risk affects the components. For example, assume you have identified a structural risk as overly ambitious plans and determine this risk affects three of the top five risks in your portfolio. You also have identified an environmental risk, in terms of whether the component will promote the organization's vision, which affects two components. Each component then has some other types of risks that affect it. From such an analysis you can see:

   a. Gaps in the portfolio
   b. Common causes
   c. Overall portfolio risk impact
   d. Rebalancing needs

10. Each portfolio risk should have a designated person as its owner. As the portfolio manager assigns the owner, the main responsibility is to:

    a. Analyze the risk for its overall impact on portfolio success
    b. Determine an appropriate response and implement it as soon as the risk occurs
    c. Monitor the situation as long as the risk is current
    d. Strive to take a negative risk and turn it into a positive opportunity

11. Your company, which has been in the blimp business for over 50 years, prides itself on its robust portfolio management process, which is especially useful now that the demand for blimps of various sizes is much greater than the capacity to produce them. One of the reasons the blimp company has been so successful over the years is it has central management of all of its resources. This approach is important:

    a. Since the various types of needed resources may be limited
    b. To ensure the right resources are assigned
    c. To monitor resource allocation
    d. As a way to assess whether existing resources have the competencies to support the components in the portfolio

12. Assume you are managing a high visibility project in your company that once it is completed will transform it into new markets and be the leader in the soft phone field. You are keeping the project a secret from external stakeholders, and you and your team have signed Non-disclosure agreements (NDAs). However, the executives and those on the Portfolio Review Board want status information on this project every two weeks. You provide it:

   a. To those on the portfolio distribution list
   b. Electronically in a format that cannot be printed or forwarded
   c. To the members of the Board and executives verbally
   d. To the portfolio manager

13. Obviously resources are more than people and include physical resources and those at the organizational level, such as the knowledge management system, helpful in portfolio management in terms of competitive intelligence. However, since most organizations operate with scarce resources, it is necessary to apply these scarce resources to support the highest ranked items in the portfolio and not to fund a proposed component unless resources are available to support it. Therefore, it is useful to determine at a collective level whether or not resources will create value greater than the cost of creating it. A portfolio report can focus on:

   a. Allocation of resources according to a Responsible, Accountable, Consulted, Inform (RACI) chart
   b. The need for drum resources and buffers to plan for their use at key times
   c. Capability and capacity
   d. Costs to acquire resources with needed knowledge, skills, and competencies

14. Your organization conducted an *OPM3* assessment focusing on its best practices in portfolio management and also on those that were needed. The *OPM3* Certified Professional felt greater attention was needed managing strategic change. This means:

   a. The project charter should be reviewed
   b. The strategic plan should be updated
   c. Budget allocation can be handled more effectively
   d. It is easier to determine which components will require specialized resources

15. As the Director of Human Resources, you were pleased to be asked to join the Portfolio Steering Committee in your organization as you feel you can contribute given the scarce subject matter experts as your company begins to enter a new market in children's toys rather than its main focus on video games and apps. Children's toys will represent a new portfolio in the company. However, while you will concentrate primarily on the resource allocation issue as you work to determine the types of projects and programs to pursue, it is important as well to recognize that:

   a. External procurements may be needed to acquire the SMEs
   b. The budgets allocated for day-to-day activities may be impacted
   c. The existing portfolio's prioritization criteria will definitely require changes
   d. The benefits from existing programs under way may not be realized

16. Assume you completed your portfolio performance management plan, and it was approved by the Oversight Committee. This plan, with an overall purpose to maximize portfolio performance, describes resource allocation and resource-related issues among other items. In it a key component is benefits realization. This emphasis on benefits planning provides:

   a. Examples for templates for benefits realization planning
   b. Methods to evaluate the expected net benefits
   c. An approach focusing on continuous improvement in overall performance
   d. Methods to assist in benefit identification

17. Assume your company recently diversified, and in addition to producing its recognized brand of ice cream products, it now also is producing cereal and nutritional products. Assume you have been using the Efficient Frontier to manage portfolio value. With diversification to these new products:

   a. Each product line should have its own portfolio to use the Efficient Frontier approach effectively
   b. The best possible portfolios are shown above the portfolio curve
   c. The same expected return from the portfolio may be possible
   d. The new potential portfolio outcomes concerning success criteria can be determined

18. Working as the portfolio manager for your business unit of a major aero-space organization means you have a variety of programs, projects, and operational activities under way. You have set up a number of reports on the progress of the portfolio for your various stakeholders, but the best approach is to monitor the progress of the portfolio against:

    a. Organizational strategy
    b. Organizational goals
    c. Specific key performance indicators for the business unit
    d. Organizational critical success factors

19. In your web app company, the portfolio is constantly changing. It is not unusual for a new proposal to be submitted each day and for other components to be terminated as a competitor was first to market. The Portfolio Review Board usually meets daily in this fast-paced environment as it:

    a. Evaluates the portfolio for specific actions it needs to take
    b. Ensures there are no open issues from past meetings that affect different components
    c. Provides a high-level view of the portfolio's direction
    d. Addresses organizational strategy

20. Your State Governor is fiscally conservative and has limited significantly the financial resources to be provided to each University in the eight Universities in the State's system. Recognizing your University is going to have a 55% reduction in its budget, your Chancellor is re-evaluating all the work that is under way to see how much it can do with fewer resources. Every department will have layoffs at all levels. However, the University does have a portfolio management process in place, which people support. This means for effective portfolio management:

    a. Each portfolio in the University should have the same restrictions in terms of available funding
    b. An effective resource assignment process is required
    c. Pareto analysis should be used to focus attention on those components with the greatest impact
    d. Portfolio maturity should be a factor in determining plans and decisions

21. One of the ten underlying principles of portfolio management involves a strategic focus. Assume you are going to have a short meeting with your CEO tomorrow, and you want to succinctly describe it. You will tell the CEO it is important since it:

    a. Emphasizes the need for portfolio management to attain strategic objectives
    b. Provides a clear basis for decision making
    c. Includes processes and change initiatives to accomplish organizational strategies
    d. Balances conflicting demands

22. Assume you are managing the corporate portfolio for your company noted for many products primarily focused on farm equipment. Recently it has diversified into other markets especially with the economic downturn in the country. While many of the traditional products are in the portfolio as new features are added to enhance customer satisfaction, one of the new product lines is a high-profile program that is ranked number five in the corporate portfolio. As the portfolio manager, you know this program has several interdependencies with other projects and programs, and recently this high-ranked program has experienced difficulties as needed technology is not available externally, and internal staff lacks the needed competencies to develop it. Its termination will be discussed at Friday's Portfolio Board meeting. You plan to discuss these interdependencies as part of your responsibilities in:

    a. Resource allocation
    b. Portfolio balancing
    c. Financial management
    d. Risk management

23. Assume you are working in the Joint Forces Command in your country. In this agency, all of the defense agencies are consolidated for better collaboration to support troops working around the world. Each program or project tends to have more than one sponsor, and since each Command is represented in this agency, the same is true for the portfolio. These sponsors are listed in the:

    a. Governance plan
    b. Portfolio charter
    c. Portfolio performance plan
    d. Portfolio Review Board charter

24. Portfolio balancing can be done in several different dimensions based on organizational preferences. When your software development company, which is CMMI Level 5 certified, began to focus on portfolio management four years ago, you started with a simplified ranking approach and now moved into using an automated, sophisticated weighted scoring software tool throughout the organization. In terms of portfolio balancing, it is appropriate to:

    a. Balance the portfolio across the organization
    b. Balance the portfolio according to categories
    c. Balance the portfolio by business unit
    d. Balance the portfolio in terms of expected value of benefits

25. While a variety of prioritization approaches are available and are useful in portfolio management and many software packages support the various approaches, it is important to recognize:

    a. Resource constraints
    b. Mandatory criteria
    c. Allocation of funds across categories based on business value
    d. Methods to determine which components should receive the highest priority

26. Assume you work in new product development, and you believe you have identified a component that will be a breakthrough for the company. However, you performed a capacity analysis with the help of your EPMO to assess resource availability especially in certain skill sets. You learned that key computer scientists required by this component were in short supply, yet you still believe this component should be in the portfolio. To convince the Portfolio Review Board to consider it, you decide to use:

    a. Resource smoothing
    b. Business value analysis
    c. Market analysis
    d. Options analysis

27. Assume you are the portfolio manager for a company that specializes in software, including portfolio management software. It has many components under way to enhance the existing product line but also to move the company into Cloud computing. You regularly prepare reports on the portfolio status but lately have had a large number of stakeholders request ad hoc reports. You decided to survey your stakeholders to learn about their information needs. You next decided to hold some one-on-one interviews with several interested and influential stakeholders in terms of communications requirements. From these interviews you are concerned that some stakeholder groups may be missing so you decided to:

   a. Conduct another survey
   b. Hold some lessons learned sessions
   c. Have a brainstorming session
   d. Convene a focus group

28. Assume you are the portfolio manager for a legacy software company. For many years, your company was one of the top five leaders in software development, but as newer and more efficient software was invented, it began to lose market share. Your company then found its services were needed as legacy systems were converted, especially since Cloud computing now is so popular. But it has lost revenues increasingly over the years. To gain market share and provide greater portfolio value, the executive team decided it should:

   a. Focus on channel partnerships
   b. Hire people with competencies in Cloud computing and enter this market
   c. Recognize change takes time but retrain employees to enhance customer satisfaction
   d. Focus on supplier value by partnering agreements

29. Your CEO was fired because of a decline in the company's profits by the Board of Directors. They have now hired a new CEO, who plans to re-shape the portfolio and has changed the company's strategic goals and objectives. The new CEO will continue the existing product line of soap products that the company has manufactured for the past 50 years but now will manufacture new products to focus on the baby boomer generation as they retire but desire to maintain a youthful appearance. It also will offer other products to new high school and college graduates who want to appear older. As the portfolio manager you should:

   a. Determine the overall impact to the portfolio performance
   b. Determine investment requirements to move to these markets
   c. Assess the competencies of the existing staff to support these new products
   d. Evaluate whether the new products can be outsourced to reduce time to market

30. Your dry foods company is faced with new regulations that dramatically change what is to be included in each product to put on redesigned food labels. The objective of the regulations is to help reduce obesity in the citizens in your country so they are aware of trans-fat food. You must be in complete compliance with these regulations in six months. As the portfolio manager you must document how you will address these regulations in a:

   a. Detailed report to the Portfolio Review Board
   b. Meeting with all employees as some existing components will be deferred to meet the requirements
   c. Portfolio performance plan
   d. Portfolio communications plan

31. Being a portfolio manager, you realize that defining value differs among organizations based on the type of organization and its strategic goals and objectives. However, you know a value measurement framework is helpful as it:

   a. Compares expected value across components
   b. Shows value in terms of tangible benefits
   c. Indicates how to best weight and score a component to authorize it
   d. Sets a baseline for a component's expected value

32. Assume you are creating a roadmap for your portfolio and will present it to key stakeholders and then to the Portfolio Review Board. You realize you will be adding additional detail to it, but you also believe its graphical format will be useful. In developing it, you decide to reference prioritization, dependencies, and organizational areas so you should consult the:

    a. Organization's strategic plan
    b. Portfolio strategic plan
    c. Portfolio
    d. Portfolio management plan

33. Assume you are holding interviews with internal and external stakeholders and subject matter experts to obtain their views on the probability that certain risks may occur and also the effect on the portfolio's objectives if the risks do occur—either positive or negative. As you conduct these interviews, a best practice to follow is to:

    a. Ask questions concerning potential outcomes concerning the portfolio success criteria
    b. Determine whether the interviewees feel rebalancing is needed to balance portfolio risks
    c. Document any assumptions
    d. Also compare recent portfolio changes

34. After the second shutdown of the Government, your Agency Administrator realized that some essential programs had to continue even during the shutdown, some existing work along with some programs and projects in the pipeline perhaps were not needed, and resources may require reallocation. This example shows the:

    a. Need to reconsider portfolio selection criteria
    b. Need to revise the portfolio mix
    c. Importance of regular reviews by the Portfolio Review Committee
    d. Need to reevaluate the entire portfolio management cycle

35. Assume you are a member of your company's Portfolio Review Board. Your Board meets quarterly to determine which new components to undertake and selects them even if it means the portfolio then will require rebalancing. As you consider the proposed business case for a component and assess the suggestions of the other Board members, a key factor is:

    a. The depth of the proposal in terms of identification of key benefits
    b. Total available resources
    c. Overall stakeholder interest
    d. Component feasibility studies

36. In preparing your communications matrix, you identified five communication areas. One is portfolio governance decisions. A communication vehicle for these decisions is:

    a. E-mails
    b. Scorecards
    c. Internal portal
    d. PMO repository

37. Assume you are putting together for the Portfolio Review Board several options for consideration of potential components and current components. You are using an approach with different probabilities to determine outcomes and EMV. Which of the following would you recommend realizing Components A and B are new, while C and D are in progress:

	Probability	Component A		Component B		Component C		Component D	
		Outcome	EMV	Outcome	EMV	Outcome	EMV	Outcome	EMV
1	50%	$15,000	$7,500	$13,000	$6,500	$20,000	$10,000	$10,000	$5,000
2	30%	$17,000	$5,100	$15,000	$4,500	$12,000	$3,600	$8,500	$2,250
3	20%	$20,000	$4,000	$15,000	$3,000	$10,000	$2,000	$5,000	$1,000

    a. Component A
    b. Component B
    c. Component C
    d. Component D

38. While the portfolio management plan is essential to ensure both effective and efficient portfolio management, when managing strategic change, it is incumbent on the portfolio manager to:

    a. Focus on its escalation policies as issues arise
    b. Ensure there is a defined process for change control and management in it
    c. Define the key roles and responsibilities of stakeholders
    d. Reassess and update it if needed

39. Assume you are responsible for portfolio management in your organization. You are responsible for managing the value of the portfolio and for recommending changes to your Portfolio Review Board to enhance its value. To do so, you monitor benefits, interdependencies between components, changes, and responsibilities and accountabilities as stated in the:

    a. Portfolio charter
    b. Portfolio management plan
    c. Portfolio performance plan
    d. Portfolio strategic plan

40. As you work to determine which of four possible components to optimize the portfolio, assume you are using the internal rate of return as the key criterion to make your recommendation. Only one new component can be added based on financial constraints. Each of the four potential components has benefits that support the strategic plan. The following are the available data:

Project A IRR	Program A IRR	Program B IRR	Project B IRR
42%	40%	36%	33%

    You recommend:

    a. Project A
    b. Program A
    c. Program B
    d. Project B

41. Assume you are working in a division in your country's Department of Interior. The Department is set up in Bureaus, and your work falls within Natural Resources. Your division is the Water Resource Division. You are responsible as the portfolio manager for the work in this Division. As you work on the portfolio for the upcoming year, you point out to the members of the Portfolio Review Committee that:

    a. The projects in the portfolio have interdependencies between them
    b. Your portfolio reflects your Division's objectives
    c. Each program and project in the portfolio have related objectives
    d. Your portfolio addresses different strategies than those in other parts of the Department

42. Assume you are the portfolio manager for a public sector organization, and it has been part of a public-private partnership for three years for highway projects. You are making recommendations as to the next program to undertake. The head of your Highway Department in your State is questioning whether the partnership is the best approach or whether it is best to work on its own. You asked the Marketing manager for assistance, and she prepared a value-for-money analysis. This approach is useful in that it:

    a. Enables an apples-to-apples comparison of the two approaches
    b. Provides a real options approach
    c. Supports a value-to-organizational vision approach
    d. Computes the expected monetary value of the two approaches

43. Because of the ongoing and iterative nature of portfolio management, the processes in it are continually repeated as new components are added, and others are completed or terminated. Revisions are constant given complexity, risks, and the rate of change. As you work to optimize the portfolio, it is helpful to:

    a. Assign components to predefined categories
    b. Prepare a flowchart
    c. Organize ideas from stakeholders into logical groupings
    d. Perform a structure analysis of roles and responsibilities

44. A useful guideline to identify the portfolio and sub-portfolios is the:

    a. Portfolio charter
    b. Portfolio performance plan
    c. Portfolio structure
    d. Roadmap

45. Programs and projects in your company, one of the largest banks in the world, are required to submit metrics as to their individual progress each month. To simplify the collection and reporting process, you held interviews with members of the Portfolio Review Board to see their areas of greatest interest and also with program and project managers to determine how difficult it would be to collect the data. You then selected 10 possible metrics to the Board, with a goal that five would be regularly reported. It is important to note that:

    a. Quantitative metrics are preferable
    b. The value is realized when components are used
    c. Customer satisfaction is the most important goal
    d. If components have interdependencies with other components, their metrics should be reported as a group

46. Your organization, a leading restaurant focusing on pancakes, is seeking to expand its portfolio. It is interested in ensuring new components support: return on investment, customer satisfaction, reputation enhancement, and branding. These four areas represent:

    a. Metrics
    b. Organizational value areas
    c. Organizational strategic objectives
    d. Critical success factors

47. Your health care firm's goals focus on products and services to provide the best care available to people in your country and has offices located throughout it. For years, to advance this goal, scientists have been pursuing research and development projects they felt were breakthroughs in the field, and many of them were successful. But, resources often were constrained in the process, and at times, some people were not fully committed. After two years of declining profits, the company has implemented portfolio management to ensure these R&D projects and other programs and projects are not pursued randomly in order that resources can be available with the required quantity and quality. This resource allocation is the responsibility of:

    a. The Enterprise PMO
    b. Portfolio governance
    c. Portfolio manager
    d. Portfolio manager working in conjunction with the HR Director

48. Assume your consulting company tried portfolio management in the past, but it was not embraced. Instead, people received bonuses if they were able to acquire new work regardless if it fit the company's strategic plan. However, the company was sold, and the new executive team asked you to be the portfolio manager. You explained it did not work in the past, but the new team has pointed out while a lot of work was won competitively, much of it was for small dollar amounts, and resources are misallocated. The new approach is to focus on business value, which has as its goal to:

    a. Achieve the greatest return on investment to the organization
    b. Maximize productivity and increase overall customer satisfaction
    c. Deliver the maximum value aligned with strategic objectives
    d. Focus on those opportunities that have the greatest likelihood of successful completion

49. In identifying risks to then manage and control, as the portfolio manager you are consulting organizational process assets such as:

    a. Commercial data bases
    b. Lessons learned
    c. Knowledge bases
    d. Values

50. Over the years, your organization has grown significantly as it has entered new markets while maintaining its presence in its traditional product line of security systems. The company now has eight different business units rather than three, which was the case only two years ago, and it set up funding originally such that it was only allocated to one business unit and could not be transferred to others. At the recently held Portfolio Oversight Committee meeting, five business units did not add components, but some were completed. The other three added a number of programs and projects, which were authorized. Now funding for these new components is an issue. This means:

    a. Another Committee meeting is required to focus on the funding problem
    b. The sponsors of the newly authorized components need to work with their business units to determine how funds will be allocated
    c. The three business units need to evaluate their portfolios and recommend termination of some components to the Committee
    d. Changes are required as to how funds are allocated

51. The members of your Portfolio Review Board and other key stakeholders tend to be risk adverse as the company has survived recent recessions and is profitable. However, in an upcoming meeting with the corporate Board of Directors, they have asked you to show the frequency of meeting certain cost objectives at various percent points. For example assume the portfolio is to meet a $41,000 target in the next month, to be 75% confident this will occur, a forecast of $50,000 is needed. This means you need to show:

    a. The needed contingency reserve
    b. The probability of achieving portfolio objectives
    c. The confidence of meeting success criteria
    d. The values of KPIs with their confidence levels

52. It is rare for organizational leaders to have an in-depth knowledge of all the work under way in the portfolio, but it is needed for portfolio decision making. If you were asked to prepare such an inventory, it would:

    a. Require one-on-one interviews to ensure all work being done was revealed
    b. Be helpful to have a statement in writing from the CEO to describe why the inventory is important
    c. Serve as the starting point for the portfolio
    d. Require assistance and support from the EPMO

53. Assume you are the portfolio manager for your training company. It decided to implement portfolio management in a major way to ensure it remained competitive in the changing market and could offer a variety of methods to deliver courses rather than only in a face-to-face setting. The company set up a Portfolio Management Group, and you are responsible for providing information on portfolio status and then providing information to those interested stakeholders about the Board's decisions. You want to make sure the reports meet stakeholder requirements. After performing a detailed communications requirements analysis, you found it interesting that stakeholders wanted information about:

    a. Portfolio infrastructure costs
    b. Goal achievement
    c. Benefit realization
    d. Changes in the roadmap

54. It is critical in portfolio management to focus on 'doing the right work'. This means stakeholder expectations and effective management of these expectations are essential. The primary conduit between the component managers and the other portfolio stakeholders is the:

    a. Program or project sponsor
    b. Portfolio manager
    c. Chairperson of the Portfolio Review Board
    d. Secretary of the Portfolio Review Board

55. Values assist in guiding actions, evaluations, and decisions. Assume your organization is considering entering into a consortium to produce a helium-controlled car. Once the helium is supplied, additional amounts will not be needed. The car is to be personally appealing with a focus on an inexpensive cost to increase marketability. The consortium will enable each firm to capitalize on the expertise of the other firms in it, but the customer will view it as a separate entity. If your organization enters into the consortium, it must justify the value to the portfolio of doing so. In addition to ensure benefits are realized a focus is needed on:

    a. Organizational value
    b. Sustainable value
    c. Managerial value
    d. Employee value

56. You are working to optimize your portfolio and determine a priority list of components to pursue. In your product development company, of the triple constraints, quality and scope dominate. This does not imply that schedule and budget are not important, but since the company requires regulatory approval for its products, quality dominates the company. Quality goals that are too low may lead to end-user dissatisfaction; however, goals that are too high may be too costly to the company. Therefore it is important to consider:

    a. Market analysis
    b. The value proposition
    c. Cash-flow requirements
    d. Risk analysis and assessment

57. Assume you are new to your organization and you were hired specifically to help implement portfolio management in your new manufacturing company. Having worked in portfolio management for the past five years, you know it is a major culture change. It is definitely a challenge at your new company because:

    a. The organization basically has many operational activities, and only a few projects are under way
    b. The organization lacks a defined strategy
    c. The few projects that are under way are not interdependent
    d. Operational activities have continued without any major changes for years

58. Assume before you prepared your portfolio management plan for your company that you did some benchmarking and learned that if you used elicitation techniques it was useful in the portfolio development stage and before there were significant scope changes because of strategy changes to the portfolio. You decided to involve the Portfolio Review Board members, other key stakeholders, and some subject matter experts in this process and then decided to poll input from the group as a majority vote. This meant you were using:

   a. Collaboration techniques
   b. Facilitation techniques
   c. Interviews and observations
   d. Negotiation techniques

59. Assume you are the corporate portfolio manager for your global organization. There is one portfolio at the corporate level, but other portfolios supporting business units and core areas of the company. One of these portfolios involves manufacturing, and its number one program in terms of priorities is to implement an enterprise resource planning system. Since it is the number one ranked program in this portfolio, it is of interest at the corporate level, and you and your team provide reports on its progress monthly. You can see that:

   a. Since earned value is being used, at this point you report the ERP system will not meet its cost and schedule goals
   b. Extensive training will be needed after the program is complete, and an infrastructure does not exist to support the ERP system
   c. Use of the ERP vendor has been underestimated, and a business case will be needed for increased funding
   d. Inadequate, up-front financial planning was done when the business case was approved

60. Your probability and impact assessment work is complete, and you are using the results to prepare the portfolio risk management plan. As you do so, it also is useful to:

   a. Define the assurance levels of the risk and its performance measures
   b. Validate with your stakeholders that your analysis meets their expectations
   c. Communicate the results with others in the organization for greater transparency
   d. Identify specific trends for each risk using qualitative and quantitative analysis

61. Assume you have been asked to prepare the portfolio management plan as you are on the staff of your company's Chief Portfolio Officer. In this plan, you will describe the different methods or approaches that your company will use to manage different types of components in the portfolio as specified in the:

    a. Strategic plan
    b. Governance model
    c. Roadmap
    d. Charter

62. Your goal as a portfolio manager is to develop a strong communications management plan to keep interested stakeholders informed about your progress in portfolio management. Although you have reached out to numerous stakeholders, you know other portfolio processes also can help in this process such as:

    a. Strategy
    b. Finance
    c. Governance
    d. Performance

63. Working to ensure the portfolio management process is one that is followed and is embraced has been a major challenge. As the portfolio manager, assume you set up meetings with the Portfolio Governance Group bi-weekly since there is constant change in your telecom company. You also want the portfolio process to be transparent. To do so, a useful tool to communicate status is:

    a. Reports on funding decisions
    b. A governance decision register
    c. The portfolio roadmap
    d. Key portfolio milestones

64. Assume you recently were hired to be the first portfolio manager at the leading producer of soft drink beverages. While the company is well known for one product, it wants to be the leader in other products and services as well. When you were hired, you told the CEO it would take time to fully implement portfolio management, and you first would prepare a number of artifacts. He set up a Portfolio Review Board, and its members are reviewing your work. As they reviewed the portfolio management plan, one of the members suggested you prepare an in-depth portfolio performance plan. Your first step is to:

   a. Align strategic management to the goals and objectives
   b. Review the prioritization model
   c. Assess the risk profile
   d. Review portfolio goals

65. You are the portfolio manager for your military-vehicle service firm, which has been in existence for 20 years. You have a number of components under way, and others in the pipeline. One component involves a new gas detection system, which uses new technology. It has interfaces with two other existing components plus one in development. Recently, a simulator, used by three components, had to be shut down completely as it was leaking nitrogen and could lead to asphyxiation. You reported it immediately to the Hazardous Materials and Pipeline Safety Administration. This is an example of a:

   a. Execution risk
   b. Structural risk
   c. Critical incident
   d. Known unknown

66. Having worked in portfolio management before, you are pleased you were selected to implement it and be the portfolio manager for your motorcycle company, well known throughout the world. As it is a new function, you worked with the Enterprise Program Management Office to ensure you had a complete inventory of the work in progress. Now as you prepared your various portfolio management artifacts, and have a Portfolio Review Board meeting upcoming in two weeks, sponsors will be proposing new components. This means:

    a. A master schedule of resource allocation is needed
    b. The meeting also should focus on reviewing existing components to see if they are aligned with current strategy
    c. People throughout the organization should know about this meeting and its decisions
    d. The meeting should have a set agenda, and each member should be contacted before it to learn of key issues

67. Assume you are working in the Joint Forces Command in your country. In this agency, all of the defense agencies are consolidated for better collaboration to support troops working around the world. Each program or project tends to have more than one sponsor, and since each Command is represented in this agency, the same is true for the portfolio. These key and major stakeholders are listed in the:

    a. Governance plan
    b. Portfolio charter
    c. Portfolio performance plan
    d. Portfolio Review Board charter

68. Your health insurance company has set up its portfolio into five different categories: research and development, IT, Medicare, government health insurance, and non-government health insurance. Funding is allocated yearly to each of these six categories. As the portfolio manager at the enterprise level, you:

    a. Ensure such allocations are reflected in the portfolio's strategic plan
    b. Meet with the CFO and determine these allocations when the budget for the fiscal year is being prepared
    c. Meet with the managers of the five portfolios once the budget allocations are known
    d. Use your existing inventory of components in the portfolio and in the pipeline to determine funding allocations

69. As the Director of the Portfolio Management Office at your worldwide furniture store, you prepare a series of reports on the status of the portfolio. One report that you use is a bubble diagram. In using it in terms of resource supply and demand, you should structure it to show:

    a. Required resources and available resources
    b. Resource availability and life cycle phase
    c. Resource competency and component probability of success
    d. Resource importance and probability of success

70. Your organization is considered a leader in knowledge management and has a Chief Knowledge Officer reporting to the CEO. It also implemented portfolio management eight years ago. As the organization focuses on continuous improvement and transformational leadership, it had an external consultant review its portfolio artifacts and do some benchmarking. One of the consultant's recommendations was to update the portfolio risk management plan since the company is embracing new and complex technology in much of its work. In updating this plan, it was useful to:

    a. Review lessons learned
    b. Determine relevant confidence limits of risk metrics
    c. Prioritize how risks are identified and listed in the risk register
    d. Determine the time in which risks are likely to have the greatest impact

71. Working to manage portfolio value is a continuous task. In doing so, as the portfolio manager, you review the monthly and any ad hoc reports submitted by component managers. This month you saw there was an excellent opportunity for major cost savings in two components in the top five on the portfolio list; however, to realize this cost reduction, these components require resources to be reallocated from other components in progress for six months. These forecasts then:

    a. Should be verified by independent estimators for accuracy
    b. Require validation by the CFO and his staff
    c. Should be accompanied by an analysis of earned value data to ensure the components are using the same method of reporting
    d. Are recommended for consideration by the Portfolio Review Board

72. As vision is the desired end state, it requires specific strategies to attain it. These strategies are best achieved by establishing:

    a. Outcomes
    b. Key performance indicators
    c. Critical success factors
    d. Goals

73. Assume it was your suggestion to the executive team as the newly appointed Chief Financial Officer for your organization to implement portfolio management. While someone has been identified to be the portfolio manager, you are developing the charter and the structure. In doing so, guidance is provided by the:

    a. Portfolio strategic plan
    b. Organization's strategic plan
    c. Portfolio roadmap
    d. Plans, policies, and documentation of stakeholder expectations

74. With the increasing use of drywall, your company, which has been in the plaster business for over three generations, is finding it harder to maintain a share of the market and to achieve a positive return on its investments. Three years ago, the corporate executives implemented a portfolio process, and they serve as the Portfolio Governance Council. They meet monthly, and after each meeting, you prepare a report of their decisions. This report is:

    a. Sent to all employees in the company as it focuses on employee empowerment and involvement
    b. Distributed only to the Governance Council to serve as a record of their meetings
    c. Is used to authorize the portfolio
    d. Is used to analyze the effects of their decisions on the company's portfolio

75. Assume you are working to ensure your organization has a balanced portfolio. You have decided to use a bubble diagram and have set it up to show the components in terms of: ease of execution [difficult or easy] and component importance [high or low]. In such an approach, bubbles are used to:

    a. Visualize components
    b. Frame the balancing problem
    c. Provide scores as outputs
    d. Focus on existing components

76. One of your goals as a portfolio manager is to ensure that your stakeholders receive the information they need for decision making. To help manage the portfolio information that is provided, you decide to gather information by holding portfolio component review meetings. Your purpose in holding these meetings is to:

    a. Ensure the components can provide the data required for status reports
    b. Use them to introduce the portfolio management information system
    c. Validate data that now are in the reports
    d. Work with component managers to plan dashboard reports

77. With the introduction of new legislation in your company, anyone now is entitled to medical services regardless of whether or not they are employed or have any pre-existing health conditions. Your insurance company's executives have been tracking this legislation as it means significant changes for your company; many employers who obtained insurance through your company may go elsewhere for lower costs. Recognizing this legislation may lead to a loss of revenue, your company decided to merge with another insurance firm to obtain greater market share. This merger, though, means some existing projects may not be needed, and the workforce will be reduced by 20 percent, Such a significant change will impact how components are categorized in your portfolio leading to:

    a. The need for a re-constituted oversight group
    b. Portfolio rebalancing
    c. A requirement to update the portfolio management plan
    d. A new portfolio prioritization model

78. Your company has had a portfolio management process in place for five years at the enterprise level, in its business units, and even in its complex programs. This year the company's Center for Excellence received an award for its work in this area from PMI®. As the Portfolio Manager, you have pre-defined metrics in place, which are critical because:

    a. They show transparency at all levels
    b. They provide stakeholders with critical information on the health of the portfolio
    c. They show the link of each initiative to the company's strategic goals
    d. They provide insight into the processes being used

79. Moving from project management to program management and now being appointed as the first portfolio manager in your cyber warfare company, you know you always wondered what happened to the various reports you had to prepare, and the metrics you had to collect. You are working now to determine critical metrics for portfolio management and decided to involve as many people as possible through questionnaires and surveys. You also held some focus groups. The purpose is to:

    a. Ensure the metrics that are collected support the SMART principle
    b. Maximize portfolio value
    c. Represent the vital few rather than the trivial many
    d. Can be gathered with minimal disruption

80. While a variety of prioritization approaches are available and are useful in portfolio management and many software packages support the various approaches, it is important to recognize:

    a. Resource constraints
    b. Mandatory criteria
    c. Allocation of funds across categories based on business value
    d. Methods to determine which components should receive the highest priority

81. Assume you are the portfolio manager for your HVAC (Heating, Ventilating, and Air Conditioning) company, one of the largest in the world. Preparing for a meeting with the Portfolio Governance Committee, you have been reviewing the success of components that have been completed as well as the progress of current portfolio components. In many cases people who only purchased heating units in the northern part of the country, and people who purchased only air conditioners in the south, now are buying state-of-the art products to easily switch as needed. You found the risks of climate change led to the need for these new energy efficient products and did so by:

    a. Sensitivity analysis
    b. Ranking and scoring techniques
    c. Investment choices
    d. Trend analysis

82. Stakeholders in your company are skeptical of the changes that will occur as portfolio management is being implemented. While some are supportive, most are not as they believe their work may be terminated. As the portfolio manager, you prepared a list of stakeholders, which is included in the:

    a. Portfolio strategic plan
    b. Portfolio charter
    c. Portfolio performance plan
    d. Portfolio management plan

83. By setting up portfolio categories and using a pair-wise comparison approach to rank components, as the portfolio manager, you feel that you are finally setting up and getting people to follow standard portfolio practices. Since portfolio management still is relatively new, progress is under way. As some components are added, and others are not continued, you are making sure if a component is terminated that it does not have dependencies with others in the portfolio. You need to then:

    a. Revise the ranking model
    b. Inform all stakeholders
    c. Update the roadmap
    d. Upgrade to a more detailed scoring model that includes dependencies with components

84. The portfolio management process ensures the components are aligned to goals. However, it is driven by:

    a. Viability
    b. Value and benefits
    c. Organizational strategy and objectives
    d. Interdependencies and resource constraints

85. Assume you recently took a seminar on portfolio management, and after you returned, you made the business case for it to the President of your consulting firm, recognizing the need to improve the capture ratio of responses to Requests for Proposals. You have been asked to implement portfolio management and recognize that you need to develop orientation and training sessions on it so everyone in the firm realizes why it is essential to pursue. To do so, the best approach is to:

    a. Have the training vendor from the seminar you attended submit a proposal to develop several training approaches
    b. Develop the courses yourself working with the vendor's materials plus other books and standards on portfolio management
    c. Work with the human resources department and have an instructional design person develop the courses
    d. Ask the PMO to develop and deliver the courses

86. Assume your food additive company performed a capacity analysis and found some resources had not maintained their skill sets and basically were not as productive as others in the company. Rather than have a massive reorganization, instead the executives decided to eliminate the jobs of these staff members, many of whom had been in the company for more than 20 years. Morale among the existing staff is low as people fear there will be more layoffs. Plus the government issued a new regulation that requires an additional Food and Drug Administration quality check before a new additive can be submitted for regulatory approval. One member of the executive team wants to acquire another company to enhance market share, and the existing plants in the Asia Pacific region require infrastructure upgrades. Given resource shortages, only one component can be selected to be added to the portfolio. The Board should select:

    a. Component A—to enhance employee morale
    b. Component B—to add staff to work with the FDA trained in quality management
    c. Component C—to acquire the competitor to increase market share
    d. Component D—to upgrade the AP's plant infrastructure

87. In your diversified chicken products company, your portfolio of components in progress consists of approximately 175 programs, projects, and other work. Generally, at each Portfolio Review Board meeting, about 35 new proposals are reviewed to see if they should be part of the portfolio. As the portfolio manager, you have set up categories for these components. They are useful to facilitate portfolio optimization because:

    a. They use filtering to eliminate certain components from consideration
    b. They help identify the components that meet requirements for consideration
    c. They serve as key evaluation criteria
    d. They address organizational strategy and objectives

88. Recognizing that different components can have different types of risks, you decide to see how each risk affects the components. For example, assume you have identified an execution as how change is managed and determine this risk affects three of the top five risks in your portfolio. You also have identified an internal risk, in terms of whether the component will promote the organization's integrity, which affects two components. Each component then has some other types of risks that affect it. From such an analysis you can see:

    a. Gaps in the portfolio
    b. Common causes
    c. Overall portfolio risk impact
    d. Rebalancing needs

89. While the Governance Board has a variety of significant roles in portfolio management, especially in terms of the recommendations it makes, these recommendations are extremely complex when they involve:

    a. Interdependencies between components
    b. Resolution of issues and risks
    c. Portfolio balancing and prioritization
    d. Resource reallocation

90. Assume the organization's strategy has undergone a significant change, and as a result the mix of components in the portfolio also will change. As the portfolio manager, you need to update your charter in order to reflect:

    a. The new 'to be' vision
    b. Interdependencies between the new components
    c. Risk tolerances
    d. Key stakeholders

91. Working as a portfolio manager in the Water Resources Department of the U.S. Geological Survey, you are following a scorecard approach to report progress to your executives on the components in your portfolio. You submit the scorecards monthly, and based on their results, your executives decide if a Portfolio Review Board meeting should be held. Your emphasis in these reports is to:

    a. Chart progress toward strategic goals and objectives
    b. Measure performance against targets and thresholds
    c. Display raw data in a visual graph
    d. Display data using a traffic light approach

92. Assume you are managing your city's portfolio, and its overall strategic goal is to promote economic development to attract more visitors to the city. It is a difficult challenge as the city is not a major metropolitan area and also is not a preferred winter or summer destination. Nonetheless, you are planning and allocating resources according to the city's strategy. Not to be overlooked as you do so is the need to:

    a. Obtain support from your key stakeholders
    b. Determine a communication strategy to explain your approach
    c. Maximize return considering the city's risk tolerance
    d. Continually update the portfolio inventory

93. As you work to determine which stakeholders had the highest degree of influence over the portfolio, you wanted to especially know about the members of the Portfolio Governance Board because:

    a. They would have numerous interrelationships with other stakeholders
    b. They would be best suited to work with people who were not portfolio management supporters
    c. The governance processes affect information requirements
    d. All of the portfolio changes, risks, and issues would be of interest to them

94. Embracing a management-by-projects culture means there tends to be far more projects to pursue than available resources. An approach then is required to guide decisions as to components in the portfolio. A best practice to follow is to:

    a. Set forth in the portfolio strategic plan a prioritization model
    b. Develop a portfolio roadmap
    c. Focus on both internal and external environmental changes
    d. Focus on sustainment of project benefits

95. You are the portfolio manager for a large county that comprises much of a major city in your country. The city also has a portfolio manager, and often you meet to discuss proposed initiatives to see if there are any dependencies. In your county, you established an approach to evaluate portfolio components to make judgments regarding their alignment and priority. In doing so, which of the following was especially helpful:

    a. Portfolio strategic plan
    b. Portfolio management plan
    c. Portfolio roadmap
    d. Portfolio charter

96. Your company's water treatment center serves three cities. Tests are conducted each day to see if the water is safe to drink. On Thursday, E coli were found in the water, but the water authority did not notify the citizens in the three cities. Instead, the citizens learned on Friday, the water was now safe to drink. The citizens are in an uproar, and your company is to blame. It is obvious one of the problems is the aging infrastructure and limited resources. To best optimize the portfolio, your CEO asked you to:

   a. Reallocate financial and other resources to new components to avoid this problem in the future
   b. First determine why citizens were not aware of the problem
   c. Analyze the testing methods in use for effectiveness
   d. Analyze the physical needs

97. Your Portfolio Review Board is scheduled to meet in a week. Resources only are available to support one project, and detailed business cases have been prepared for two of them. Your company has a policy of being risk adverse. Consider the following table:

	Project A	Project B
Benefits	$750,000	$25,000
Costs	$500,000	$15,000

   Which project would you recommend to the Board, and what else would you mention to them?

   a. Project A and it has less risk associated with it
   b. Project B and it has less risk associated with it
   c. Project A as the benefits will be realized in a shorter time period
   d. Project B but other qualitative items are not available

98. As you prepare a list of possible components for your railroad to consider since most of its programs and projects are large and complex, your management team has suggested in your analysis of which components to pursue that you conduct statistical simulations of budgets, schedules, and resource allocations. You therefore decide to use:

   a. Net present value
   b. Decision trees
   c. Monte Carlo analysis
   d. An interrelationship diagraph

99. You have a portfolio component that is using earned value analysis. It is at the 15% point of completion, and it is evident that it cannot be completed as planned. Adding resources will not solve the problem, and at the last Portfolio Review Board meeting, the Board members decided to terminate this component based on its various risks. They then decided the resources allocated to this component could be transferred to other portfolio components enhancing their early completion and avoiding risks from competitors. As the portfolio manager, you:

   a. Worked with the component managers to ease the transition
   b. Documented these decisions in portfolio reports
   c. Set up both quantitative and qualitative metrics to determine the usefulness of adding resources to the other components
   d. Met with the affected component managers and their teams to explain these changes

100. Assume you are the portfolio manager for the Federal Railroad Administration. Funding is provided annually according to the Government's budget process. Any monies that are not spent at the end of a fiscal year are lost, and there are some restrictions in place concerning whether funds can be transferred to different programs, projects, or operations work in the agency. Therefore, to maximize the use of funding you require:

   a. Regular reports on funds for authorized components
   b. Projections on a quarterly basis as to the funds components require
   c. Mechanisms for internal audits to ensure funds are allocated effectively
   d. Accurate estimates of the funds needed when the component is proposed for consideration

101. Although you are the Portfolio Manager in your company, you also are the Enterprise Program Office Director, and it consumes most of your time. Your executive team has lost interest in regular portfolio review meetings, the few that are held are routine, and any proposed component receives automatic approval. The Executive Vice President for Human Resources spoke to the CEO as she noticed some people seem to be completely overloaded, they are actively looking for new opportunities, and are then leaving the company. She also pointed out that others seem to have idle time. Her comments got the CEO's attention, and you have been asked in your role as the portfolio manager to create an up-to-date list of qualified components in the portfolio. You are doing so in order to:

    a. Have an accurate understanding as to how resources are allocated
    b. Determine how best to use resource leveling on approved components
    c. Identify, categorize, score, and rank components
    d. Review the initial business cases for the components and assess their validity in terms of strategic goals

102. Assume you are working to prepare the low-level schedule and timelines for portfolio components. You want to make sure, as the portfolio manager, for your country's initiatives to promote an awareness of the importance of climate change, that each component then can be monitored and tracked to assess performance. To do so, you should:

    a. Set up KPIs for each component that are consistent for ease of measurement
    b. Determine the critical success factors at the portfolio level and then ensure each component contributes toward their realization
    c. Review the portfolio roadmap
    d. Review the portfolio performance plan

103. Specific types of communication technology that are used such as communication media, record retention policies, and security information are examples of:

    a. Organizational process assets
    b. Portfolio process assets
    c. Items in the information distribution process
    d. Items needed to cover the portfolio communications management plan

104. Assume your Department can only pursue one of the five possible components listed below:

Pairwise Comparison Example

1	2						
1	3	2	3				
1	4	2	4				
1	5	2	5	3	5	4	5

Based on the above data, you should select:

a. Component 5
b. Component 3
c. Component 2
d. Component 1

105. After a recent Portfolio Review Board, the portfolio was optimized, and some components were added, while others were removed. Various portfolio reports also require updates such as:

a. Affected organization areas
b. High-level time frame
c. Budget approvals or exceptions
d. Value/benefits

106. You have been a successful program manager for many years in your State Department of Agriculture. During this time, you managed large programs, and some had major risks to mitigate especially in the information systems area as new software would be released that was commercially available, and you knew it would then enhance the benefits to your customers if you acquired it. You were the first in the Department to get your PfMP, and it led to a promotion to become the first portfolio manager. After a year in this position, you find managing risks and issues to be totally different because:

a. You must focus attention on external, political risks
b. You are concentrating more on long-term initiatives
c. Your focus is on determining the risk tolerances of stakeholders, both internal and external
d. You emphasize strategic fitness of the portfolio

107. The Manage Portfolio Value Process, while ongoing, has proved to be successful as you work to implement portfolio management. Within six months, you were able to show the usefulness of a simple scoring model to the Oversight Group, and they requested a more sophisticated approach in which weights could be assigned to criteria. This shows:

   a. An acceptance of portfolio management in the organization
   b. The usefulness of the portfolio roadmap
   c. A link between using scoring models and benefits analysis
   d. The importance of documenting lessons learned

108. Assume you are the portfolio manager for your cereal company, which has diversified its product line significantly in the last two years to keep up with its leading competitor located in a different state. Your executive management team learned the other cereal company had implemented portfolio management from a contractor and believes it is essential since the economy is struggling, and resources are constrained. Your first step has been to identify the existing operational work, projects, and programs as well as to learn about proposed components of the portfolio. This list:

   a. Was easy to obtain as you used what was available from the Enterprise Program Management Office
   b. Is part of the portfolio strategic plan
   c. Was gathered through interviews with people from each business unit
   d. Was prepared through questionnaires and the use of cross-functional focus groups

109. Assume you are a functional manager in your medical device company in research and development. Your scientists have determined a new product that will be a breakthrough for the company, and you want to serve as the sponsor for this component and present it to your Portfolio Review Board. You will need resources from other parts of the company to commercialize it. As you prepare your proposal you are following the key descriptors set up by the portfolio staff and will include:

   a. Risk reduction
   b. Regulatory and compliance issues
   c. Internal and external dependencies
   d. Qualitative benefits

110. Assume you work for a technology company that is publically owned, and the value of its stock is tracked daily by the CFO and is reported to the portfolio manager. Quarterly meetings are held with stockholders as the company went public through an Initial public offering (IPO) last year. These stockholders:

    a. Have different communications requirements than other stakeholders
    b. Typically receive information as to the portfolio health before each meeting
    c. Want to attend all Portfolio Review Board meetings
    d. Are considered external stakeholders

111. Working to best optimize resource supply and demand in your telecom company, authorized components are prioritized to help in resource allocation. This means as you work to do so, you need to review the portfolio because:

    a. It provides guidance in terms of recommendations if there are changes in strategy and resource availability
    b. Resource requirements are balanced according to the resource pool
    c. Components are not authorized unless resources are available to support them
    d. Sponsors assess resource requirements and their availability before proposing a component to be in the portfolio

112. Working to prepare the portfolio performance management plan, assume you have been involving others in the process to help secure their later support of the plan. You also reviewed historical information and other artifacts. Not to be overlooked is a/an:

    a. Benefit schedule
    b. Organization chart
    c. Regulatory requirements
    d. Governance model

113. For years, your aerospace company has been a leader in the development of sophisticated avionic hardware systems around the world. The executives want to continue with this well-recognized product line, but also they decided it is time to move into state-of-the art software to complement the hardware products. You have been asked to assess whether there are skill set limitations in the company to assess resource capacity internally. To do so, you decide to:

    a. Focus on needed competencies and develop competency profiles for the internal staff
    b. Interview staff members based on performance evaluations to determine their interest in the new product line
    c. Work with Human Resources and review the education backgrounds of internal staff and see if people have taken recent training
    d. Set up a contingency plan by asking the Procurement Department to issue a Request for Information to external consulting firms

114. You are the CIO of a real estate investment trust (REIT) that invests in apartments and condominiums in more than 50% of the states in your country. Your organization has as its goal to respond to any concerns that arise within 24 hours; for example, you want to make sure Wi-Fi sites are operational if there are any power outages, and people have soft phone service available 24/7. You are a member of the REIT's Portfolio Review Board, and as a member of the executive team in terms of portfolio risk management, you want to focus on:

    a. Issues with product support
    b. Identifying and managing liabilities
    c. Interaction of component risks
    d. Inconsistent processes

115. Assume at this point as the portfolio manager in your mattress company, you are activating portfolio components, updating portfolio reports, and documenting the decisions made at the recent Portfolio Governance Committee meeting as the company moves into new markets. You are therefore:

    a. Communicating decisions to all stakeholders
    b. Maintaining a decision log from the Committee meeting
    c. Updating the portfolio strategic plan
    d. Authorizing the portfolio

116. As the portfolio manager, you must engage stakeholders and build and maintain outstanding relationships with them as much as possible. After identifying and classifying them, you want to make sure you provide the specific information each stakeholder group requires. Therefore you prepare:

    a. A communications strategy matrix
    b. A communications matrix
    c. A reporting frequency matrix
    d. A stakeholder matrix

117. Assume your bank has decided to implement portfolio management. It is starting with the business unit you lead, which is responsible for new products, and then will set the stage to implement it throughout the bank, including at the enterprise level. You have executive support and commitment to implement it in your business unit. A key first step is to:

    a. Set up a governance structure
    b. Define roles and responsibilities for implementation
    c. Prepare a portfolio performance plan
    d. Prioritize the work to be done

118. Assume you are the portfolio manager for your pork producing company, the market leader in your country. Over time, the industry has recovered from trichinosis as a risk. Your company has added new components to its portfolio, and many have been to demonstrate to the public that its products are safe. It implemented the Agriculture Department's and Food and Drug Administration's Hazard Analysis and Critical Control Point (HACCP) regulations and is enhancing its image as 'the other white meat'. However, now the entire industry is faced with a new epidemic known as porcine epidemic virus, which is affecting pigs in 22 different states, and profits have decreased significantly. New components now must be added to the portfolio. This situation shows:

    a. Resource re-allocation is required
    b. Risk management is essential
    c. The ROI of the new components must be determined
    d. Portfolio rebalancing has led to the new components being in the top five priority list

119. Assume as the portfolio manager you have conducted a stakeholder analysis, gap analysis, and a readiness analysis as your cereal company is now entering the ice cream market. A team was formed and located off site to determine whether this market was one in which your company could compete, and its recommendation to do so was accepted by the Portfolio Review Board. Now you need to:

    a. Acquire resources to support the new line of ice cream products
    b. Change the prioritization model
    c. Develop a communications strategy for use internally and also externally
    d. Set completely new performance metrics for all products

120. Working to monitor the portfolio especially in terms of its value to the organization, you had each component manager prepare monthly variance reports. Of the components ranked in the top 10, six of them realized they would not require some of their initial funding and still would be completed as planned. This means:

    a. Three-point estimating should be used as funds are allocated
    b. Next year's budget can be adjusted
    c. Historical data would be useful on estimates versus actual costs
    d. The amount of contingency and management reserves can be decreased

121. As you prepare your communications management plan you realize that stakeholder requirements may change over time, and these changes then will need to be reflected in updates to the plan. As the plan defines the overall communication process not only for gathering information but also in determining recipients of it, it also:

    a. Sets expectations for the project team members
    b. Describes how and when to effectively conduct lessons learned meetings
    c. Ensures continued confidence by all stakeholder groups as to their importance
    d. Sets expectations for effective stakeholder engagement

122. In your telecom company, a number of criteria must be considered as you develop your approach to prioritize components in the portfolio. Your management insists that to be competitive the products must be first to market or the window of opportunity is lost with the result being not only lost revenues but also lost productivity. Another criterion to consider is:

    a. External dependencies
    b. Goals and objectives
    c. Customers
    d. Regulatory compliance

123. Different types of risks affect the portfolio, and they may be positive or negative. As the portfolio manager, one has to maximize the opportunities and minimize the threats. An example of a negative portfolio risk is:

    a. External participants who are highly specialized
    b. Integrated systems
    c. A large number of concurrent programs and projects
    d. Full-cost estimates for programs and projects

124. In your architectural organization, each program or project requires some specialized subject matter experts at certain time frames. Because of the interdependencies between components, often these SMEs are needed at the same time. Assume since this is a critical issue in the company, it invested in resource planning and allocating software, and dashboards can be prepared. The goal is to use these dashboard reports to:

    a. Determine whether to use external consultants
    b. Assess specific costs for additional budget
    c. Ensure everyone has access to the software for ease of communication
    d. Assist in scheduling adjustments

125. Having worked in portfolio management for several years, assume you were hired as the portfolio manager for a Real Estate Investment Trust, one of the largest in your country that specializes in apartments. The company continues to grow and wants to maximize value and profits for its investors. As you set up processes and procedures for portfolio management, you know from past experience that buy in from executives is insufficient. As you prepare a communications strategy, you focus on:

    a. Satisfying important information needs of stakeholders
    b. Surveying stakeholders through a questionnaire to determine information requirements
    c. Using focus groups to assist in determining information needs
    d. Focusing first on the executive team's communications requirements and then involving others

126. Many people in your country are no longer eating food from cans because of the risk of botulism and eColi O1H747. Your low acid canned foods company is seeing its revenues decrease as a result, and it is updating its overall strategy for the company to diversify into other markets as well as to add an aggressive marketing campaign to ensure the public that its low acid canned foods are generally recognized as safe by the Food and Drug Administration. This means in terms of portfolio management:

    a. Each proposed component must demonstrate business value before it is undertaken
    b. The ROI for existing components should be reviewed to determine if they should continue
    c. The existing inventory of work should be validated against the updated strategy
    d. The benefits to be realized by existing components require standard KPIs tied to critical success factors

127. Different stakeholders will have different portfolio reporting requirements. Sponsors for example will have a great interest in:

    a. If the portfolio will meet organizational strategy
    b. Status in achieving benefits
    c. Overall portfolio value
    d. Status in terms of other authorized components

128. As the portfolio manager in the third largest automotive manufacturer in your country, you have a large number of components especially new vehicles each year but also support for dealers, advertising, maintaining the brand image, increasing market share plus continuous improvement initiatives. You have contingency reserve to use to prepare to handle any risks that may occur, which is based on:

    a. Expected monetary value
    b. Return on investment
    c. Expected financial benefits
    d. Equity protection

129. Although it has taken significant time, you and your team inventoried all the work under way in your new product development company. This list of components should be:

    a. Included in the portfolio roadmap
    b. Part of the portfolio management plan
    c. Prioritized for effective resource allocation
    d. Maintained by the portfolio manager and continually updated

130. Assume you have determined the prioritization criteria your Portfolio Review Board will use, and you have reviewed the criteria with your key stakeholders to attain their buy off and occurrence. The purpose in establishing these criteria is to:

    a. Ensure each component in the portfolio is in alignment to strategic goals
    b. Incorporate the key stakeholders' risk tolerances as a criterion for consideration
    c. Enable comparison among components
    d. Set forth measurable goals with KPIs

131. Capability and capacity analysis are useful tools in portfolio performance management. In using this type of analysis a best practice is to:

    a. Employ a resource management process
    b. Evaluate knowledge, skills, and competencies
    c. Use it once portfolio resources are included in the portfolio performance management plan
    d. Evaluate resource optimization

132. Assume your company is a leading producer of AA and AAA batteries. However, it is a competitive market, and customers desire batteries with a longer life and a smaller size so they will not require replacement. As you set up categories for portfolio components, you will continue to produce your current product line as well as pursue advanced products to meet customer needs. A useful component category, therefore, is:

    a. Benefits
    b. Business imperatives
    c. Stakeholders
    d. Technology capabilities

133. Assume you are the portfolio manager for a leading drug store in your country that offers numerous products. In the past four years, nearly every store has had to enlarge its pharmacy unit and hire additional staff members with the aging population. Observing this change, two years ago, stores set up clinics to provide customers with immediate care. As you see the growth in the stores in the health arena, you are looking at trends and realize:

    a. Alcohol, tobacco, and sugar soft drink products should no longer be offered
    b. Each store requires a balance between its health care services and products that may have adverse health effects
    c. Customers wonder if they should trust the health care services offered given the other available products
    d. For the health care clinics to be viewed with integrity, a medical doctor must be available at each store

134. Assume you are co-owner of a small consulting firm. Previously, you worked as a managing partner in one of the larger consulting firms in your country that had a defined portfolio management process to determine key opportunities to pursue to focus not solely on proposal win ratio but to aggressively emphasize capture ratio. Now in your new company in terms of portfolio management, the best practice to follow is to:

    a. Work with your business partner in terms of portfolio management
    b. Have your Board of Directors serve as a Portfolio Review Board
    c. Involve your business partner plus the firm's subject matter experts in portfolio decisions
    d. Set up an independent group of advisors to meet quarterly as a Portfolio Review Board

135. Your IT company has been successful as it is able to deliver projects on time without the need for rework and within the allocated budget. Your customers have been astonished with the results and are using your company for additional work, plus they have been recommending your company to others. Your company is experiencing tremendous growth and wants to ensure it can take on the new work with existing resources, both people and systems, or whether it will need to use outsourcing. Given its outstanding reputation, your executives wish to avoid the need to outsource. You have been asked to perform a capacity analysis. A best practice is to:

    a. Prepare a model of the current configuration and modify it to determine future capacity requirements

    b. Determine and document existing assumptions

    c. Inventory staff members to assess their level of competencies and existing workload

    d. Use resource leveling in an enterprise project management information system

136. Your organization tried implementing portfolio management in the past, but even though it purchased a sophisticated automated software system for scoring and prioritization, it was not successful. After a year of complaints about the system, the executive team disbanded it. Now, resources are scarce, and some executives who were involved in the earlier attempt have left the company. The new CEO asked for a list of ongoing programs, projects, and operational work, and you could not provide one even though you direct the Project Management Office. He wants this inventory so it then can be used as the starting point to implement portfolio management. You and several others stated it did not work previously and do not believe it will work in the company. The CEO then hired a person to be the portfolio manager from outside who reports directly to him. The portfolio manager is emphasizing the importance of resources working on initiatives aligned with organizational strategy as part of the:

    a. Portfolio performance plan

    b. Portfolio roadmap

    c. Portfolio communications strategy

    d. Portfolio charter

137. Before any information in your web-based technology company is communicated externally, it must be submitted to the company's Public Relations Department to ensure sensitive information is not disclosed inadvertently to competitors. The Public Relations Director must sign off on all external information. You need to, as the portfolio manager:

    a. Make sure your team is aware of this requirement
    b. Meet with this Director to explain the purpose of the communication
    c. Limit the frequency of external communications to stockholders
    d. State this requirement in the communication plan

138. Assume you are preparing the first portfolio risk management plan for your outsourcing company, which typically handles call centers around the world. While the company has implemented portfolio management and has a Portfolio Oversight Group, it did not previously assess risks to the portfolio itself. Instead, it assumed risks would be managed at the project level. However, numerous customer complaints have been received. The root cause is once a new call center is established, limited if any planning is done as the manager rushes to have it ready and operational as soon as possible. This has led to a lack of understanding as to what is required for the call centers to be successful. In preparing this plan, you are reviewing the portfolio management plan because it:

    a. Contains the portfolio vision statement
    b. Provides the organization's risk tolerance
    c. Provides guidance on stakeholder engagement
    d. Includes the portfolio performance matrices

139. Information and direction about the organization's vision, mission, prioritization, and resources should be obtained before the portfolio's strategic plan is developed by reviewing:

    a. Knowledge repositories
    b. Portfolio roadmap
    c. Organizational process assets
    d. Governance model

140. Roadmaps may be prepared to show different elements, and at the beginning they may not provide details of the various components. As a high-level plan at the portfolio level, the roadmap:

    a. Identifies internal and external dependencies
    b. Serves as a master schedule to show the timing of approved components
    c. Contains all the details of program and project roadmaps
    d. Serves to identify issues

141. As you are the portfolio manager for your state government agency, which is undergoing a series of budget cuts, you are focusing attention on managing risks to the portfolio as the budget is reduced. You realize in this process the time and budget for risk management also will be reduced; these data are in the:

   a. Portfolio performance plan
   b. Portfolio strategic plan
   c. Portfolio management plan
   d. Portfolio financial plan

142. As the portfolio manager you have worked to consider the complexities involved of the interdependencies in your programs, projects, and ongoing work. As your portfolio process has been implemented for three years, a key challenge is that senior executives tend to change priorities often even though programs and projects are being implemented. Often these programs and projects are cross-functional, and the result is your process is not coordinated. You recognize there is a need to change, and you met with the Chair of the Portfolio Review Board and have her support for a transparent approach for portfolio standards and prioritization. You should:

   a. Revise the balancing process
   b. Revise the portfolio management plan
   c. Revise the scoring model
   d. Set up a standard method to communicate change

143. Assume your telecom company is time constrained and needs to be first to market with new smart phones with features that are different from those of the competition and also have the traditional features desired by your existing customers. The Portfolio Review Board meets weekly to assess performance and to consider new components. Lack of technical resources is a recurring issue. To make the case for acquiring new resources, you decided to assess capability and capacity. This approach is:

   a. Part of the PMIS
   b. Included in finite capacity planning and reporting
   c. Used to prepare a detailed forecast of ongoing and future capability needs
   d. Used to identify resource capacity and capability

144. Since you work for a global aerospace and defense organization, it decided to pilot the implementation of portfolio management in its cargo aviation business unit. You were asked to lead this initiative and realized before you could proceed, you should find out information about all the existing projects, programs, and operational activities. This task took three months to complete, but with this list, you now can use it to set up categories for the work that is under way and to also define criteria to use to propose new components. As a result you should:

a. Determine how to optimize the existing work
b. Ensure the criteria and categories are aligned with the portfolio roadmap
c. Set up a Portfolio Review Board
d. Prepare a portfolio performance plan

145. You prepared a portfolio risk management plan when you replaced the previous portfolio manager three years ago. However, recent structural and execution risks have affected the portfolio adversely, resulting in lost opportunities and a decrease in overall return on investment. You are updating the risk management plan as now stakeholders can see its value. In doing so, you can use some portfolio process assets such as:

a. Lessons learned
b. Portfolio algorithms
c. Vision statements
d. Risk categories

146. You are working diligently to ensure people throughout the organization realize the importance of portfolio management. To do so, you are preparing a communications strategy. When you distribute it, you will be able to show you can satisfy information requirements in order to:

a. Show the data you plan to collect will be analyzed
b. Provide credibility for a portfolio management process
c. Meet the organization's objectives
d. Push information to stakeholders on portfolio status

147. After three months, you have a list of all the program, project, and ongoing work being done in your 500-person Division of your State Government Agency. With this list, the next step is to:

a. Determine the prioritization model to follow
b. Convene a meeting of the Portfolio Review Board
c. Assess gaps in meeting the Agency's strategic objectives
d. Prepare a portfolio performance plan

148. Even though you do not work in an industry that is heavily regulated by your government, such as in new product development, health, or safety, recently your government issued a mandate that senior management of all corporations must certify the accuracy of reported financial statements to prevent any accounting fraud. These controls are to be implemented in the next fiscal year. This obviously is not in your portfolio strategic or management plans but is an example of a(n):

a. Mandated component
b. Enterprise environmental factor
c. Strategic change
d. Emergent program

149. As soon as you complete the portfolio risk management plan, and you have been working on it now with a team, you realize you need to update some organizational process assets such as:

a. Risk checklists
b. Risk register
c. Lessons learned
d. Risk interview guide

150. Consider the following situation:

Project/Program	Cost	IRR	Risk	Type
Program A	$30 million	21%	0	Operational improvement
Program B	$30 million	22%	1	Operational improvement
Program C	$30 million	26%	2	Capital expansion
Project D	$30 million	24%	1	New product
Project E	$30 million	22%	2	New product
Project F	$30 million	28%	3	New strategy

Assume you have been asked to perform a prioritization analysis based on these data. You realize risk is a major concern to the company, but you have some data available about potential benefits. These data show A and D have the greatest benefits. A and D are followed in terms of benefits by C, then B, then F, and finally E. Assume three of the programs and projects can be added to the portfolio when the Board meets. Your recommendation is to select:

a. A, B, and C
b. A, D, and C
c. A, F, and C
d. D, B, and C

151. Assuming a portfolio manager position means one has more stakeholders than in program, project, or operational roles. The goal is to identify all interested stakeholders but often overlooked are:

   a. Consumer groups
   b. Alliances
   c. Associations
   d. External resource providers

152. Assume after the acquisition of the natural gas transmission company by your company, a natural gas distribution company, was approved by the various regulatory agencies. You now are overseeing more components with this acquisition as the portfolio manager. While you had each of the components in your company set up in various categories, this approach had not been followed by the transmission company. You explained to its portfolio manager and staff such an approach enables:

   a. Common criteria for portfolio optimization
   b. A similar approach to track contribution to strategic goals
   c. A way to set up a common set of decision filters
   d. An alignment with the prioritization model

153. Working previously in the financial industry and studying finance and risk in graduate school, you are familiar with Markowitz's Efficient Frontier theory. Now assume you are the portfolio manager for a state government agency. Your agency has a reputation of being risk adverse but given recent budget cuts, you have convinced your executive team it needs to pursue some new programs and projects to demonstrate its benefits to the state. You decided to apply the Efficient Frontier concepts to show them the current state of its components in terms of risk and associated costs. You explained the portfolio is efficient if it has:

   a. A mix of components—from high risk/high return to low risk/low return
   b. The ability to quantify the value of risk in monetary terms
   c. The possible overall portfolio value with the greatest possible benefits
   d. The best possible expected level of return for its level of risk

154. Your organization has a defined portfolio management process that it has followed for three years. As the portfolio manager, you keep your various plans up to date, and because of numerous regulatory changes involving the telecom industry, it is time to review and update the risk management plan. You have several key stakeholders working with you as you realize the importance of this plan in maintaining a competitive advantage. One way you and your team are assessing the various risks that may impact the structure of the portfolio is to use:

    a. Portfolio component charts
    b. Portfolio reports
    c. Weighted ranking and scoring techniques
    d. Risk metrics

155. Assume your pork producing company finds that there is an over-abundance of pork products and competitors in the marketplace even though it has had to implement Hazard Analysis and Critical Control Point (HACCP) processes that are a regulatory requirement. Profits are lower than ever before in the history of the company. Management is changing the company's strategy to also focus on seafood products. You have been asked to complete a gap analysis to:

    a. Determine resource capacity
    b. Assess risks with this change
    c. Compare the current portfolio mix with that with this change
    d. Determine any requirements that must be addressed before the change is implemented

156. Your online ordering company wants to add a component to its portfolio that its sponsor believes will outdistance the competition, but it has risks and also will be subject to regulatory approval. The purpose is to use parachutes to deliver the merchandise ordered through small helicopters so the recipients receive their orders within three hours of the on line purchase. As the portfolio manager you recognize this component is a major change and will require resources if it is approved. You are now performing change management using a change structure that:

    a. Requires a change request
    b. Facilitates impact analysis
    c. Needs to assess dependencies
    d. Requires an update to the roadmap

157. Assume you work in a weak matrix structure in your pharmaceutical company in which most of the program and project managers are coordinators, and most of the staff that supports them are in functional organizations. On some high priority programs, staff may be dedicated to the program full time for a short time period; however, operational work often takes precedence especially in manufacturing. The demand for some of the pharmaceutical products often outpaces the available supply, and shelf life is short. These fluctuations of resources then:

    a. Require use of resource soothing
    b. Led to the development of resource heuristics as to how best to manage the portfolio
    c. Require sign-offs from functional managers on the portfolio charter concerning resource availability
    d. Impact the availability of the resources for the work managed within the portfolio.

158. Review the following graphic. Assume now your portfolio is only 12% likely to meet is target of $41,000. Your Portfolio Review Board is dissatisfied in your management of the value of the overall portfolio. You explain the current mix of components is too risk adverse, and additional investment is required. The Board Chair then wants the needed investment to have a 75% likelihood, and you state it is:

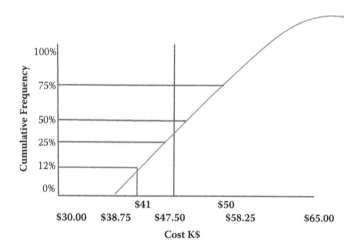

    a. $100,000
    b. $50,000
    c. $125,000
    d. $65,000

159. Assume you are the portfolio manager for a telecommunications company. Your company was about to launch a new and easy to use smart phone with more features than any existing phones on the market at a lower price. However, although the phone was due to market in five days, the Federal Communications Commission issued today a regulation that would make your new phone not available for use in airplanes. Thus additional work must be done, and your executives are wondering whether a new phone should be developed for this new feature. You are ensuring that if a new phone is developed, or if the almost completed product is not to be marketed, there is still alignment to the organization's strategy. As you complete an analysis of alternatives, you also should ensure results of the analysis are reflected in the:

    a. Benefits realization plan
    b. Portfolio process assets
    c. Portfolio roadmap
    d. Portfolio performance plan

160. Working to prepare the communications plan, a best practice to follow is to use the roadmap. By doing so, it:

    a. Shows the overall portfolio timeline, useful for determining the frequency of reporting
    b. Provides information about interdependencies that may affect objectives
    c. Emphasizes milestones and the timing of key benefits
    d. Shows applicable constraints

161. Because your company's Portfolio Review Board consists of the Directors of its five business units and is chaired by the CEO, the meetings tend to be contentious as there is limited funding available to authorize all the proposed programs and projects. Dissension also is the norm if resources are reallocated from one business unit to another. As a result, the CEO:

    a. Strives to use consensus to make decisions, but this approach rarely is effective
    b. Decided to use an outside facilitator when meetings are held
    c. Uses multi-voting and makes the final decision
    d. Often uses a Delphi approach

162. As you focus on managing the value of the portfolio, you find that portfolio variance/alert reports are helpful. Assume you have been using a 'traffic light' format as it is easy to prepare, but an objective is to:

    a. Add in blue to show completed components
    b. Show dependencies between components with a different color
    c. Set it up to show the organizational value areas in the company
    d. Use a standardized format across components

163. You want to ensure that the Portfolio Review Board is able to make key decisions at each meeting. As the portfolio manager, you and your staff are responsible for scheduling the meetings, providing the agenda, taking minutes, tracking open issues, and documenting and communicating decisions that are made to key stakeholders. Before each meeting, you feel it is a best practice to:

    a. Evaluate if the benefits of the portfolio are aligned with organizational strategy
    b. Provide information about the status of each component of the portfolio
    c. Use a balanced scorecard approach to show contribution to strategy
    d. Provide a 'traffic light' approach to show components by category

164. One key artifact to review as the portfolio communications management plan is prepared is the:

    a. Portfolio management plan as it shows all elements in it have communications requirements
    b. Portfolio performance plan as it sets forth needed reports and their frequency
    c. Portfolio strategic plan since it shows the need for strategic alignment
    d. Portfolio benefits realization plan to determine reports on progress in benefit realization, transition, and sustainment

165. Each year, you update the portfolio roadmap so people within the organization can see component status, interdependencies, constraints, and business value, among other things. This year, however, two major programs in the portfolio were cancelled as they were government contracts, and the government lacked funds to complete them. Your management then had to reduce staffing. These two programs had dependencies with other components in the portfolio. This means that:

    a. The other components may need to be cancelled
    b. The proposed benefits from the other components require analysis to see if they can be realized and sustained
    c. The value measurement criteria for portfolio components require updates
    d. The overall value of the portfolio is affected adversely

166. While your organization maintains a decision register after each meeting of the Portfolio Review Board, this register only notes decisions that are made when a component is added to the portfolio or if the Board terminates a component. If the component is terminated, the reason for the termination is not listed. Also if a proposed component is not approved, the reason is not listed. It is evident this register is lacking in its usefulness, which means it requires updating as it is:

a. In the portfolio management plan
b. A portfolio process asset
c. An organizational process asset
d. Part of the governance model

167. While there are a number of recommended contents of the portfolio strategic plan, a guiding principle is to:

a. Document assumptions and constraints
b. Recognize stakeholder risk tolerances
c. Recognize the portfolio will evolve through progressive elaboration
d. Define the portfolio vision and objectives to align with organizational strategy

168. Various people are responsible for communications to different stakeholder groups, both internal and external to the organization. These delegations of authority are:

a. Organizational process assets
b. Contained in the portfolio performance plan
c. A section in the portfolio communication management plan
d. Portfolio process assets

169. Each time the Portfolio Governance Group meets the goal is to review the existing components and any that are proposed to ensure the portfolio has the best mix to attain strategic objectives. As the portfolio manager, you find these meetings, if facilitated accordingly, are effective decision-making sessions. However, you tend to have open issues after every meeting. These open issues:

a. Should be tracked in an issue register
b. Are managed as described in the charter
c. Are considered portfolio process assets
d. Require an owner to manage them until they are closed

170. After the stakeholder analysis is complete, a best practice is to put stakeholders into a matrix to develop a communications management strategy. A simple but useful approach is to set it up to show:

   a. Level of authority and level of interest
   b. Level of authority and level of involvement
   c. Level of influence and level of impact
   d. Level of influence and level of interest

Answer Sheet for Practice Test 1

1.	a	b	c	d		20.	a	b	c	d
2.	a	b	c	d		21.	a	b	c	d
3.	a	b	c	d		22.	a	b	c	d
4.	a	b	c	d		23.	a	b	c	d
5.	a	b	c	d		24.	a	b	c	d
6.	a	b	c	d		25.	a	b	c	d
7.	a	b	c	d		26.	a	b	c	d
8.	a	b	c	d		27.	a	b	c	d
9.	a	b	c	d		28.	a	b	c	d
10.	a	b	c	d		29.	a	b	c	d
11.	a	b	c	d		30.	a	b	c	d
12.	a	b	c	d		31.	a	b	c	d
13.	a	b	c	d		32.	a	b	c	d
14.	a	b	c	d		33.	a	b	c	d
15.	a	b	c	d		34.	a	b	c	d
16.	a	b	c	d		35.	a	b	c	d
17.	a	b	c	d		36.	a	b	c	d
18.	a	b	c	d		37.	a	b	c	d
19.	a	b	c	d		38.	a	b	c	d

39.	a	b	c	d		61.	a	b	c	d
40.	a	b	c	d		62.	a	b	c	d
41.	a	b	c	d		63.	a	b	c	d
42.	a	b	c	d		64.	a	b	c	d
43.	a	b	c	d		65.	a	b	c	d
44.	a	b	c	d		66.	a	b	c	d
45.	a	b	c	d		67.	a	b	c	d
46.	a	b	c	d		68.	a	b	c	d
47.	a	b	c	d		69.	a	b	c	d
48.	a	b	c	d		70.	a	b	c	d
49.	a	b	c	d		71.	a	b	c	d
50.	a	b	c	d		72.	a	b	c	d
51.	a	b	c	d		73.	a	b	c	d
52.	a	b	c	d		74.	a	b	c	d
53.	a	b	c	d		75.	a	b	c	d
54.	a	b	c	d		76.	a	b	c	d
55.	a	b	c	d		77.	a	b	c	d
56.	a	b	c	d		78.	a	b	c	d
57.	a	b	c	d		79.	a	b	c	d
58.	a	b	c	d		80.	a	b	c	d
59.	a	b	c	d		81.	a	b	c	d
60.	a	b	c	d		82.	a	b	c	d

	a	b	c	d
83.	a	b	c	d
84.	a	b	c	d
85.	a	b	c	d
86.	a	b	c	d
87.	a	b	c	d
88.	a	b	c	d
89.	a	b	c	d
90.	a	b	c	d
91.	a	b	c	d
92.	a	b	c	d
93.	a	b	c	d
94.	a	b	c	d
95.	a	b	c	d
96.	a	b	c	d
97.	a	b	c	d
98.	a	b	c	d
99.	a	b	c	d
100.	a	b	c	d
101.	a	b	c	d
102.	a	b	c	d
103.	a	b	c	d
104.	a	b	c	d

	a	b	c	d
105.	a	b	c	d
106.	a	b	c	d
107.	a	b	c	d
108.	a	b	c	d
109.	a	b	c	d
110.	a	b	c	d
111.	a	b	c	d
112.	a	b	c	d
113.	a	b	c	d
114.	a	b	c	d
115.	a	b	c	d
116.	a	b	c	d
117.	a	b	c	d
118.	a	b	c	d
119.	a	b	c	d
120.	a	b	c	d
121.	a	b	c	d
122.	a	b	c	d
123.	a	b	c	d
124.	a	b	c	d
125.	a	b	c	d
126.	a	b	c	d

127.	a	b	c	d
128.	a	b	c	d
129.	a	b	c	d
130.	a	b	c	d
131.	a	b	c	d
132.	a	b	c	d
133.	a	b	c	d
134.	a	b	c	d
135.	a	b	c	d
136.	a	b	c	d
137.	a	b	c	d
138.	a	b	c	d
139.	a	b	c	d
140.	a	b	c	d
141.	a	b	c	d
142.	a	b	c	d
143.	a	b	c	d
144.	a	b	c	d
145.	a	b	c	d
146.	a	b	c	d
147.	a	b	c	d
148.	a	b	c	d

149.	a	b	c	d
150.	a	b	c	d
151.	a	b	c	d
152.	a	b	c	d
153.	a	b	c	d
154.	a	b	c	d
155.	a	b	c	d
156.	a	b	c	d
157.	a	b	c	d
158.	a	b	c	d
159.	a	b	c	d
160.	a	b	c	d
161.	a	b	c	d
162.	a	b	c	d
163.	a	b	c	d
164.	a	b	c	d
165.	a	b	c	d
166.	a	b	c	d
167.	a	b	c	d
168.	a	b	c	d
169.	a	b	c	d
170.	a	b	c	d

Answer Key for Practice Test 1

1. c.   An executive

    Executives are interested in any risks that impact the portfolio, such as the organization's operating model, customer brand, the organization's reputation, impact on strategy and objectives, and any impact on any existing products or services.

    Portfolio Management Standard, p. 122

    Task 3 in the ECO in Risk Management

2. c.   Include portfolio benefits, results, and value expected in the portfolio strategic plan

    As the purpose of the portfolio strategic plan is to address organizational strategy and alignment to it, the benefits, or desired outcomes of actions, behaviors, products or services, along with results and value to be realized, should be documented in it so there is common understanding by stakeholders as to why portfolio management is essential.

    Portfolio Management Standard, p. 46

    Task 2 in the ECO in Strategic Alignment

3. c.   Financial standing

    The contracts management team is interested in the overall financial standing of the organization in case there are changes that affect portfolio components. They also are interested in progress of programs and projects along with contract impact and changes.

    Portfolio Management Standard, p. 111

    Task 3 in the ECO in Communications

4. b.   Considered a legitimate right

    In setting up and maintaining a governance model for portfolio management, organizational governance is a consideration. Among other things, it is a process in which the organization responds to legitimate rights of stakeholders; in this example to treat all stakeholders ethically.

    Portfolio Management Standard, p. 62

    Task 1 in the ECO in Governance

5. b. The portfolio requirements should be evaluated

As an output of the Authorize Portfolio process the portfolio management plan may require updates. These updates may involve the process to authorize portfolio components because the requirements need evaluation with the merger in this situation.

Portfolio Management Standard, p. 80

Task 3 in the ECO in Governance

6. a. The criteria to optimize the portfolio may be the same as that used in the scoring model

A variety of tools and techniques can be used to optimize the portfolio and have a list of components for possible authorization. Many organizations find it facilitates the process to use the same approach as in the scoring model.

Portfolio Management Standard, p. 73

Task 6 in the ECO in Performance

7. c. Lessons learned

Other portfolio process assets to update include portfolio-related people, processes, and technology; performance metrics; and risk management.

Portfolio Management Standard, p. 55

Task 7 in the ECO in Strategic Alignment

8. d. Options analysis

Options analysis is a variant of cost/benefit analysis. If the organization elects not to consider the component, the result is that often it is difficult or even impossible to consider it in the future. It then can pursue it when it is considered profitable when additional information is available. The component is treated as an option.

Portfolio Management Standard, pp. 74, 102

Task 7 in the ECO in Performance

9. b. Common causes

This type of analysis is portrayed in a risk component chart, a qualitative risk tool and technique used in the Manage Portfolio Risks process, and is helpful to see if there are common causes affecting various portfolio components thereby facilitating decision making in terms of next steps

Portfolio Management Standard, pp. 133, 135

Task 5 in the ECO in Risk Management

10. c. Monitor the situation as long as the risk is current

The risk register includes an identified owner for each risk. This person is responsible for managing the risk, which means analyzing it, assigning response actions, and monitoring it until it is closed. Depending on the severity of the risk in terms of probability and impact on the portfolio before implementing a response, approval by the portfolio manager and possibly the oversight group may be required.

Portfolio Management Standard, p. 130

Task 4 in the ECO in Risk Management

11. b. To ensure the right resources are assigned

Portfolio reports are an output of the Manage Supply and Demand process. Resource reports especially are reviewed to ensure the right resources are assigned to the right components at the right time.

Portfolio Management Standard, p. 96

Task 1 in the ECO in Performance

12. d. To the portfolio manager

Portfolio component reports are provided to the portfolio manager. He or she then reviews the reports and can consolidate them as appropriate.

Portfolio Management Standard, p. 116

Task 5 in the ECO in Communications

13. c. Capacity and capability

Resource capacity and capability reports demonstrate whether resources are available to do the allocated work in the portfolio and also if they have the required knowledge, skills, and competencies to do the work.

Portfolio Management Standard, p. 37

Task 8 in the ECO in Performance

14. a. The project charter should be reviewed

   The project charter is an input to the Manage Strategic Change process. It should be reviewed to see if it and the portfolio still are aligned and, if required, are updated.

   Portfolio Management Standard, p. 56

   Task 7 in the ECO in Strategic Alignment

15. b. The budgets allocated for day-to-day activities may be impacted

   Achieving portfolio objectives may impact functional groups in their day-to-day operations. For example, an operational budget may be influenced by portfolio management activities including allocation of resources.

   Portfolio Management Standard, p. 5

   Task 3 in the ECO in Communications Management

16. b. Methods to evaluate the expected net benefits

   The benefits realization planning shows expected benefits for a given portfolio. It then allows the Oversight Group the ability to evaluate the expected net benefits to help make informed decisions.

   Portfolio Management Standard, p. 91

   Task 4 in the ECO in Governance

17. c. The same expected return from the portfolio may be possible.

   Such diversification may lead to reduced risk. The efficient frontiers are not static.

   Portfolio Management Standard, p. 103

   Task 6 in the ECO in Performance

18. a. Organizational strategy

   The best approach is to monitor portfolio performance against organizational strategy

   Portfolio Management Standard, p. 9

   Task 8 in the ECO in Performance

19. d. Addresses organizational strategy

The portfolio is an input to the Provide Portfolio Oversight process as the oversight group is concerned with ensuring the current or planned components support organizational strategy.

Portfolio Management Standard, p. 82

Task 1 in the ECO in Governance

20. b. An effective resource assignment process is required

The lack of efficient and effective functional area processes and procedures may have a considerable impact on portfolio management; one of which is the lack of an effective resource assignment process to support the execution of components once approved through portfolio management.

Portfolio Management Standard, p. 27

Task 6 in the ECO in Strategic Alignment

21. c. Includes processes and change initiatives to accomplish organizational strategies

A strategic focus supports the definition of portfolio management. It includes the organizational processes and change initiatives to enable organizations to prioritize, select, evaluate, and allocate scarce resources to accomplish organizational strategies; these strategies are ones consistent with mission, vision, and values.

PMI (2013) Managing Change in Organizations, p. 46

Task 1 in the ECO in Strategic Alignment

22. d. Risk management

Risk threat management is critical especially when there are interdependencies between high profile portfolio components. The cost of the failure of a component then is significant. The same is true if the risk of one component increases the risk to other components.

Portfolio Management Standard, p. 17

Task 3 in the ECO in Risk Management

23. b. Portfolio charter

Among other things, the portfolio charter lists the portfolio sponsors and sets forth portfolio management roles and responsibilities.

Portfolio Management Standard, p. 49

Task 1 in the ECO in Governance

24. a. Balance the portfolio across the organization

    The situation shows it is a mature organization. In balancing, many balance components within the same categories and later move to address diverse concerns and strategy across the organization.

    Portfolio Management Standard, p. 71

    Task 6 in the ECO in Performance

25. b. Mandatory criteria

    Mandatory criteria include regulatory or operational requirements, which must be met. These components then are in the portfolio regardless of the prioritization approach that is used.

    Portfolio Management Standard, p. 70

    Task 5 in the ECO in Strategic Alignment

26. a. Resource smoothing

    Resource loading requirements over time, such as through the use of resource leveling or resource smoothing are an example of a quantitative analysis technique. Resource smoothing is useful to be able to adjust a model of the proposed schedule to ensure requirements for any predetermined resources do not exceed any capacity limits.

    PMBOK®, pp. 179–180

    Portfolio Management Standard, pp. 75, 95

    Task 7 in the ECO in Performance

27. c. Have a brainstorming session

    Brainstorming is useful to gather ideas from a number of participants and can help identify new stakeholder groups. These sessions also can change or confirm requirements from existing groups.

    Portfolio Management Standard, p. 112

    Task 3 in the ECO in Communications

28. a. Focus on channel partnerships

    Channel partner value is especially important in the IS/IT field. Becoming a channel partner, enables an organization to capitalize on another firm's credibility and recognition and acquire credentials, promoting increased market share and overall portfolio value.

    Portfolio Management Standard, p. 103

    Task 9 in the ECO in Performance

29. a. Determine the overall impact to the portfolio performance

Changing priorities means resources must be realigned, and the cost structure of any needed resources will impact the portfolio's performance. The portfolio manager assesses impacts to determine if they are value added or adverse.

Portfolio Management Standard, p. 118

Task 5 in the ECO in Communications

30. c. Portfolio performance plan

This plan includes a section as to how the portfolio manager will address compliance regulations.

Portfolio Management Standard, p. 63

Task 4 in the ECO in Governance

31. a. Compares expected value across components

The value measurement framework organizes the value to be realized, shows how it will be measured, and recognizes both tangible and intangible benefits.

Portfolio Management Standard, p. 96

Task 9 in the ECO in Performance

32. c. Portfolio

The charter should be reviewed as the roadmap is described. It contains information on the components as well as prioritization, dependencies, and organization areas.

Portfolio Management Standard, p. 51

Task 8 in the ECO in Strategic Alignment

33. c. Document any assumptions

Each interviewee will have various assumptions concerning probability and impact. These assumptions are a source of risk, and assumption analysis may be needed to determine their validity.

Portfolio Management Standard, p. 126

Task 2 in the ECO in Risk Management

34. b. Need to revise the portfolio mix

   Disruptions to the organization are an indication that it is time to revise the portfolio mx even though the portfolio process is continual.

   Portfolio Management Standard, p. 25

   Task 3 in the ECO in Performance

35. b. Total available resources

   Resources are limited in organizations. Capacity analysis is required to assess resource requirements including people, available finding, and assets such as physical requirements.

   Portfolio Management Standard, p. 74

   Task 5 in the ECO in Governance

36. c. Internal portal

   Since many stakeholders are interested in these decisions and since they will probably be reviewed frequently, use of an internal portal is the most appropriate communications vehicle.

   Portfolio Management Standard, p. 112

   Task 4 in the ECO in Communications

37. a. Component A

   Sensitivity analysis uses probabilities as a decision analysis method. In this example, Component A has the highest EMV and would be recommended.

   Rechenthin, p. 59

   Portfolio Management Standard, p. 48

   Task 5 in the ECO in Strategic Alignment

38. d. Reassess and update it if needed

   The portfolio management plan describes the needed oversight for the portfolio with criteria to develop, maintain, and control the portfolio and its components. It should be reassessed to ensure it is still supportive when strategic changes do occur and updated as required based on the intended strategy.

   Portfolio Management Standard, p. 53

   Task 5 in the ECO in Performance

39. b. Portfolio management plan

   The portfolio management plan is an input to the Manage Portfolio Value process. Among other things, it has a major emphasis on benefits especially in identifying and assessing their impact on objectives and ensuring components will realize them successfully.

   Portfolio Management Standard, p. 96

   Task 9 in the ECO in Performance

40. c. Project A

   Project A has teh highest IIR and thus is the most desirable to undertake.

   Milosevic, pp. 42–44

   Portfolio Management Standard, p. 74

   Task 5 in the ECO in Governance

41. d. Your portfolio addresses different strategies than those in other parts of the Department

   Organizations may have more than one portfolio each addressing organizational strategies and objectives.

   Portfolio Management Standard, p. 3

   Task 1 in the ECO in Strategic Alignment

42. a. Enables an apples-to-apples comparison of the two approaches

   This approach is used by marketing especially in non-profit or government agencies as it estimates the whole life cycle cost of carrying out the project in the traditional approach by the Highway Department as well as doing it through the partnership so the costs can be considered equally.

   Portfolio Management Standard, p. 13

   Task 2 in the ECO in Governance

43. b. Prepare a flowchart

    Flowcharts are helpful to show the entire portfolio process and when it is time to revisit the portfolio to see if it has the mix of components that best support organizational strategy. A flowchart also enables a focus on key activities. They are a tool and technique in Optimize Portfolio in probability analysis.

    Portfolio Management Standard, p. 75

    Task 7 in the ECO in Performance

44. c. Portfolio structure

    The portfolio structure is a guideline to identify the portfolio and sub-portfolios since it is based on the organizational areas, hierarchies, and goals for each component.

    Portfolio Management Standard, p. 49

    Task 4 in the ECO in Strategic Alignment

45. b. The value is realized when components are used

    The overall value of the portfolio is delivered when its components are used by its beneficiaries.

    Portfolio Management Standard, p. 86

    Task 3 in the ECO in Performance

46. b. Organizational value areas

    Values represent how the organization's leaders make decisions. By setting up value areas, decision makers can see how proposed and existing components support the values of interest.

    Portfolio Management Standard, p. 85

    Task 1 in the ECO in Strategic Alignment

47. b. Portfolio governance

    Even if a PMO is in place in the organization, portfolio governance is responsible for allocating resources at the time required, in the right quantity and with the required quality levels.

    Portfolio Management Standard, p. 62

    Task 1 in the ECO in Governance

48. c. Deliver the maximum value aligned with strategic objectives

Portfolio value is the aggregate value of the components in the portfolio. It also ensures the maximum value considers the organization's risk tolerance.

Portfolio Management Standard, p. 96

Task 8 in the ECO in Performance

49. d. Values

Values are used in decision making and are critical to portfolio management. Organizational process assets are an input to the Manage Portfolio Risks process.

Portfolio Management Standard, p. 132

Task 4 in the ECO in Risk Management

50. d. Changes are required as to how funds are allocated

Updates to portfolio process assets are an output of the Authorize Portfolio process; this situation shows there is a need to change how funds can be transferred among components.

Portfolio Management Standard, p. 80

Task 3 in the ECO in Governance

51. b. The probability of achieving portfolio objectives

Different organizations are willing to assume different levels of risk. A risk adverse organization, as described in this situation, requires a higher level of investment than one that is more risk tolerant.

Portfolio Management Standard, pp. 127–128

Task 2 in the ECO in Risk Management

52. c. Serve as the starting point for the portfolio

An inventory of the work in progress is needed as it comprises the portfolio; it is also an input to the development of the portfolio strategic plan.

Portfolio Management Standard, p. 43

Task 4 in the ECO in Strategic Alignment

53. b. Goal achievement

    Performance reports often focus on variances and goal achievement. These portfolio reports are an input to the Manage Portfolio Information process.

    Portfolio Management Standard, p. 116

    Task 5 in the ECO in Communications

54. b. Portfolio manager

    The portfolio manager receives information from the component managers and provides it to the Portfolio Review Board. He or she also provides information from the Review Board to the component managers and other stakeholders especially in terms of how the portfolio components are aligned with the organization's strategic goals, on progress, changes, and impact on components.

    Portfolio Management Standard, pp. 14–15

    Task 3 in the ECO in Communications Management

55. c. Managerial value

    If an organization is to enter into a consortium, for it to be beneficial, the member firms must have similar managerial values in order that it is viewed as a single entity to others.

    Portfolio Management Standard, p. 103

    Task 9 in the ECO in Performance

56. a. Market analysis

    Market analysis is a tool and technique in the Optimize Portfolio process. It is necessary to understand end-user expectations as this knowledge is then used to help determine whether the component should be in the portfolio. It provides information, such as market size, market segmentation, and sales potential that will increase the probability of the component's success.

    Milosevic et al, p. 368

    Portfolio Management Standard, p. 75

    Task 7 in the ECO in Performance

57. b. The organization lacks a defined strategy

The portfolio must be aligned to organizational strategy and if not, organizational leaders need to question why the work is being done. It represents the intent, direction, and progress of the organization.

Portfolio Management Standard, p. 3

Task 1 in the ECO in Strategic Alignment

58. a. Collaboration techniques

In this approach, the plan has the benefit of polling input from the team as a consensus or majority vote. These votes can be obtained electronically and may be anonymous.

Portfolio Management Standard, p. 61

Task 4 in the ECO in Governance

59. b. Extensive training will be needed after the program is complete, and an infrastructure does not exist to support the ERP system

In this situation, support will be needed from IT, and the portfolio manager is responsible for considering these requirements and ensuring the infrastructure and training needs can be met for portfolio components.

Portfolio Management Standard, pp. 13–14

Task 7 in the ECO in Performance

60. a. Define the assurance levels of each risk and its performance measures

As part of the tools and techniques in the Develop Risk Management process and as an activity in conjunction with probability and impact analysis, such confidence limits should be documented.

Portfolio Management Standard, p. 127

Task 2 in the ECO in Risk Management

61. c. Roadmap

The roadmap is used in the development of the portfolio management plan. In this example, since it shows the various components and their dependencies in a chronological manner, it then leads to different management approaches to consider.

Portfolio Management Standard, p. 60

Task 8 in the ECO in Strategic Alignment

62. d. Performance

Drawing from other portfolio management processes such as performance and risk can help in the development of the communications management plan.

Portfolio Management Standard, p. 107

Task 2 in the ECO in Communications

63. c. The portfolio roadmap

The roadmap presents a high-level timeline of the components in the portfolio and provides an integrated view; this enables it to be a useful tool for communications.

Portfolio Management Standard, pp. 82–83

Task 5 in the ECO in Governance

64. d. Review portfolio goals

In preparing the portfolio performance plan, the first step is to review the portfolio goals in the portfolio strategic plan and the objectives to reach them.

Portfolio Management Standard, p. 87

Task 1 in the ECO in Strategic Alignment

65. a. Execution risk

An execution risk refers to the ability to manage change and includes how change is managed in performing components. It includes interactions and interdependencies between risks in components in process or under development.

Portfolio Management Standard, p. 123

Task 3 in the ECO in Risk Management

66. a. A master schedule of resource allocation is needed

An inventory of portfolio resources and capabilities is needed to plan the consolidated demand to support new and existing components.

Portfolio Management Standard, p. 92

Task 7 in the ECO in Performance

67. b.   Portfolio charter

Among other things, the portfolio charter lists the key portfolio stake-holders and sets forth portfolio management roles and responsibilities.

Portfolio Management Standard, p. 49

Task 1 in the ECO in Governance

68. a.   Ensure such allocations are reflected in the portfolio's strategic plan

Allocation of funds and resources to different types of initiatives and how they contribute to the organization's objectives are included in the portfolio strategic plan to aid the decision makers especially in comparing new and existing components with each other.

Portfolio Management Standard, p. 39

Task 6 in the ECO in Strategic Alignment

69. a.   Required resources and available resources

In this approach, based originally on concepts by Cooper, Edgett, and Klienschmidt, the bubble diagram is helpful for balancing decisions. It can be set up with axes in which criteria are scored, and the portfolio components are shown in four quadrants. These charts are easy to use and can offer an unbiased view, but many different ones tend to be needed to show the various prioritization criteria.

Milosevic, pp. 74–83

Portfolio Management Standard, p. 96

Task 3 in the ECO in Performance

70. a.   Review lessons learned

Lessons learned are an organizational process asset and an input to the Develop Portfolio Risk Management Plan process. Since in this example, the organization focuses on knowledge management, lessons learned should be available and easy to access.

Portfolio Management Standard, p. 125

Task 2 in the ECO in Risk Management

71. d. Are recommended for consideration by the Portfolio Review Board

    Portfolio reports are an output of the Manage Portfolio Value process. In this situation, these cost savings and the actions recommended then require consideration from governance.

    Portfolio Management Standard, p. 104

    Task 9 in the ECO in Performance

72. d. Goals

    Organizations establish goals to move to their desired vision. Goals have objectives that then are measured, and strategies are developed to achieve goals.

    Portfolio Management Standard, p. 7

    Task 8 in the ECO in Strategic Alignment

73. a. Portfolio strategic plan

    The purpose of the portfolio strategic plan in this situation is to provide guidance as it describes the portfolio vision and objectives, benefits, key risks, assumptions, and constraints.

    Portfolio Management Standard, p. 48

    Task 1 in the ECO in Performance

74. c. Is used to authorize the portfolio

    Governance decision reports are an example of portfolio reports, an input to the Authorize Portfolio process. These reports authorize components and provide detailed information on why the component was selected.

    Portfolio Management Standard, p. 79

    Task 5 in the ECO in Governance

75. a. Visualize components

    Bubble diagram graphs are an example of a graphical analytical method used as a tool and technique in Optimize Portfolio. The bubbles represent components, both existing and proposed, according to predefined balancing criteria.

    Milosevic, pp. 74–75

    Portfolio Management Standard, pp. 75–76

    Task 7 in the ECO in Performance

76. c. Validate data that now are in the reports

These meetings are an example of an elicitation technique, used in the Manage Portfolio Information process. They serve to validate and better understand the data in component reports to ensure the data that are provided are useful to stakeholders.

Portfolio Management Standard, p. 116

Task 5 in the ECO in Communications

77. b. Portfolio rebalancing

Significant changes in strategy will impact component categorization in the portfolio or prioritization and require that the portfolio be rebalanced as a result.

Portfolio Management Standard, p. 63

Task 7 in the ECO in Strategic Alignment

78. d. They provide insight into the processes being used

Metrics are useful to drive accountability for performance improvement, to provide insight into current processes, and to determine any needed changes to these processes.

Portfolio Management Standard, p. 25

Task 3 in the ECO in Performance

79. b. Maximize portfolio value

Metrics are in the portfolio performance management plan. The objectives are to collect the metrics that best ensure resource performance and maximize portfolio value.

Portfolio Management Standard, pp. 89, 104

Task 4 in the ECO in Governance

80. b. Mandatory criteria

Mandatory criteria include regulatory or operational requirements, which must be met. These components then are in the portfolio regardless of the prioritization approach that is used.

Portfolio Management Standard, p. 70

Task 5 in the ECO in Strategic Alignment

81. d. Trend analysis

Trend analysis is an example of a tool and technique in the Manage Portfolio Risk process. By reviewing trends based on risks as in this situation it can lead to different responses and actions to take to mitigate the risks.

Portfolio Management Standard, p. 132

Task 5 in the ECO in Risk Management

82. b. Portfolio charter

A list of stakeholders is included in the charter to best ensure the portfolio will deliver business value for them by formally identifying them.

Portfolio Management Standard, p. 39

Task 1 in the ECO in Performance

83. c. Update the roadmap

As the portfolio process is defined, if components change or if dependencies change, the roadmap requires updates to it.

Portfolio Management Standard, p. 70

Task 5 in the ECO in Governance

84. c. Organizational strategy and objectives

Before the portfolio process begins, it is driven by the organizational strategy and objectives that are set by the executives.

Portfolio Management Standard, p. 21

Task 1 in the ECO in Strategic Alignment

85. d. Ask the PMO to develop and deliver the courses

One area in which the PMO can assist in portfolio management is to develop and conduct training in it and mentor people in the portfolio management skills, tools, and techniques.

Portfolio Management Standard, p. 18

Task 5 in the ECO in Communications

86. b. Component B—To add staff to work with the FDA trained to work in quality management

In considering resource scenario analysis, this company requires FDA approval for its products. Its best option is to ensure the time-to-market window is streamlined as much as possible so additional resources should be allocated to work with the FDA.

Portfolio Management Standard, p. 95

Task 7 in the ECO in Performance

87. d. They address organizational strategy and objectives

Assigning components to categories helps compare those that address similar organizational needs and strategic objectives. In this way, they facilitate optimization to ensure selected and managed components address organizational strategy and objectives.

Portfolio Management Standard, p. 68

Task 4 in the ECO in Strategic Alignment

88. b. Common causes

This type of analysis is portrayed in a risk component chart, a qualitative risk tool and technique used in the Manage Portfolio Risks process, and is helpful to see if there are common causes affecting various portfolio components thereby facilitating decision making in terms of next steps

Portfolio Management Standard, pp. 133, 135

Task 5 in the ECO in Risk Management

89. d. Resource reallocation

Recommendations by the Governance Board may include new components, changing or terminating existing components, and reallocating resources among components. It is in resource reallocation where the recommendations are complex especially if organizational constraints are in place.

Portfolio Management Standard, p. 19

Task 5 in the ECO in Governance

90. d. Key stakeholders

The charter is updated to reflect the new strategic objectives and key or major stakeholders and their communication requirements.

Portfolio Management Standard, p. 55

Task 7 in the ECO in Strategic Alignment

91. a. Chart progress toward achieving strategic goals and objectives

Scorecards and dashboards often are used interchangeably. Scorecards are visual displays used as strategically-oriented performance metrics to chart progress toward achieving strategic goals and objectives by comparing progress against targets and thresholds.

Kerzner (2011), p. 201

Portfolio Management Standard, p. 37

Task 3 in the ECO in Performance

92. c. Maximize return considering the city's risk tolerance

Each organization has a pre-defined risk profile and tolerance. In optimizing the portfolio and allocating resources according to strategy, this risk profile or tolerance must be considered to maintain portfolio return.

Portfolio Management Standard, p. 71

Task 7 in the ECO in Performance

93. c. The governance processes affect information requirements

It is necessary to know the Governance Board members and the specific governance processes to be supported by regular communications.

Portfolio Management Standard, p. 110

Task 3 in the ECO in Communications

94. a. Set forth in the portfolio strategic plan a prioritization model

Prioritization analysis is a tool and technique to use in developing the portfolio strategic plan. The prioritization model to be used then is described as part of this plan.

Portfolio Management Standard, p. 46

Task 6 in the ECO in Strategic Alignment

95. b. Portfolio management plan

The purpose of this plan is to define management's approach and intent to identify, approve, procure, prioritize, balance, manage, and report on the components in the portfolio.

Portfolio Management Standard, p. 66

Task 4 in the ECO in Governance

96. d. Analyze the physical needs

Capacity and capability analysis are more than resources alone. In this situation, an asset capability and capacity analysis is appropriate to understand the physical needs including equipment, buildings, and systems to understand any constraints that may limit existing or proposed components.

Portfolio Management Standard, p. 74

Task 7 in the ECO in Performance

97. d. Project B but other qualitative items are not available.

Based only on the results of the benefit/cost analysis, Project B is preferred as its Benefit Cost Ratio is 1.67. However, a weakness of only using benefit/cost analysis is that other qualitative items are not available, which if known may change the recommendation.

Rechenthin, p. 158

Portfolio Management Standard, p. 51

Task 7 in the ECO in Strategic Alignment

98. c. Monte Carlo analysis

Monte Carlo Analysis is another form of probability analysis, a tool and technique in the Optimize Portfolio process, in which a model is computed many times with the input values chosen at random from the probability distribution of these variables.

Portfolio Management Standard, p. 75

Task 7 in the ECO in Performance

99. b. Documented these decisions in portfolio reports

Portfolio reports are an output of the Manage Portfolio Risks process. Governance recommendations should be part of the portfolio reports as managing risks produces the need for various reports about the portfolio.

Portfolio Management Standard, pp. 134–135

Task 6 in the ECO in Risk Management

100. a. Regular reports on funds for authorized components

Once components are authorized, funds are updated regardless if annual or multi-year funding is used. Regular reports are required as an output of the Authorize Portfolio process.

Portfolio Management Standard, p. 80

Task 3 in the ECO in Governance

101. c. Identify, categorize score, and rank components

This is the purpose of the Define Portfolio process and is necessary for ongoing evaluation, selection, and prioritization.

Portfolio Management Standard, p. 64

Task 4 in the ECO in Strategic Alignment

102. c. Review the portfolio roadmap

In developing the portfolio management plan, it uses information from the roadmap's high-level strategic direction and extended timeline to create these lower-level schedules for the components.

Portfolio Management Standard, p. 60

Task 4 in the ECO in Governance

103. b. Portfolio process assets

The portfolio manager updates and adds to portfolio process assets, which include ones from stakeholders involved in the portfolio to influence its overall success. These are examples of portfolio communication requirements, one of many portfolio process assets.

Portfolio Management Standard, pp. 36–37

Task 5 in the ECO in Communications

104. c. Component 2

Each possible pair of components is compared. The table lists the choices for five components; as noted Component 1 is compared to Component 2 and then 1 and 3, etc., until all possible comparisons are made. The formula is (n X [n-1])/2) or ([5 × 4]/2). As the comparison is done, Component 2 is the recommended choice.

Rechenthin, pp. 54–55

Portfolio Management Standard, pp. 68–69

Task 5 in the ECO in Strategic Alignment

105. d. Value/benefits

As an output of the Optimize Portfolio process, portfolio reports may require updates. Of particular importance is the value of the components to the organization, and the benefits each component contributes to overall strategic goals and objectives.

Portfolio Management Standard, p. 77

Task 2 in the ECO in Performance

106. d. You emphasize strategic fitness of the portfolio

Risks at the program or project level tend to arise within a specific program or project. At the portfolio level, the emphasis is on increasing the probability and impact of opportunities and decreasing negative impacts adverse to the portfolio's value, strategic fitness, and component balance.

Portfolio Management Standard, p. 120

Task 1 in the ECO in Risk Management

107. d. The importance of documenting lessons learned

Lessons learned in portfolio management are an output of the Manage Portfolio Value process. By analyzing them, a more effective process for managing value can be determined.

Portfolio Management Standard, p. 104

Task 10 in the ECO in Performance

108. b. Is part of the portfolio strategic plan

The portfolio strategic plan, among other things, contains a portfolio structure, which is a list of all the portfolio components and other work.

Portfolio Management Standard, p. 39

Task 4 in the ECO in Strategic Alignment

109. d. Qualitative benefits

While a number of descriptors should be used, qualitative and quantitative benefits should be included to also support strategic goals.

Portfolio Management Standard, p. 67

Task 2 in the ECO in Governance

110. d. Are considered external stakeholders

Stakeholders, once identified, should be classified into those that are internal or external to the organization as a way to plan communications to them.

Portfolio Management Standard, p. 110

Task 3 in the ECO in Communications

111. b. Resource requirements are balanced according to the resource pool

The portfolio is an input to the Manage Supply and Demand process as it shows active and planned components, which represent resource requirements or demand. They then are balanced to determine how to effectively manage the portfolio with limited resources.

Portfolio Management Standard, p. 94

Task 6 in the ECO in Performance

112. a. Benefits schedule

Portfolio process assets should be reviewed as the performance management plan is prepared. In addition to various documents, they also include various schedules. A benefit schedule is useful as it will show when benefits from components are expected and thus assist in overall portfolio value.

Portfolio Management Standard, p. 88

Task 4 in the ECO in Governance

113. a. Focus on needed competencies and develop competency profiles for internal staff

The first step in this capability analysis is to determine the needed competencies to support the new product line. Then, competency profiles can be reviewed if they exist, or they can be prepared to see whether internal staff members possess the needed skill sets for the proposed work.

Portfolio Management Standard, p. 48

Task 5 in the ECO in Strategic Alignment

114. b. Identifying and managing liabilities

At the executive level, the focus often is on safeguarding stakeholder investment, company assets, and identifying and managing liabilities as risk concerns emphasize portfolio value.

Portfolio Management Standard, p. 122

Task 3 in the ECO in Risk Management

115. d. Authorizing the portfolio

This is the purpose of the Authorize Portfolio process as resources are allocated to develop or execute components, and reports are updated and distributed.

Portfolio Management Standard, p. 77

Task 1 in the ECO in Performance

116. d. A stakeholder matrix

This matrix is used for stakeholder analysis. It lists each stakeholder group and then describes their roles, interests, and expectations, providing useful information for communications planning.

Portfolio Management Standard, p. 111

Task 3 in the ECO in Communications

117. b. Define roles and responsibilities for implementation

As portfolio management is implemented, roles and responsibilities in the implementation process require definition as changing business processes is a difficult undertaking.

Portfolio Management Standard, p. 24

Task 4 in the ECO in Governance

118. b. Risk management is essential

Risk management is ongoing in portfolio management. It is useful to help to identify potential improvements in the portfolio in terms of components. In this example, the new components are essential to increase quality and customer satisfaction for the company.

Portfolio Management Standard, p. 119

Task 1 in the ECO in Risk Management

119. b. Change the prioritization model

The portfolio strategic plan will need revision because of this change, and it contains the prioritization model, among other items.

Portfolio Management Standard, p. 55

Task 7 in the ECO in Strategic Alignment

120. b. Next year's budget can be adjusted

   In this situation, the reports provide information to support adjustments to the upcoming budget. Portfolio reports are an output of the Manage Portfolio Value process.

   Portfolio Management Standard, p. 103

   Task 9 in the ECO in Performance

121. a. Sets expectations for project team members

   Setting expectations for team members is necessary concerning actions and processes necessary to facilitate critical alignment among people, ideas, and information.

   Portfolio Management Standard, p. 113

   Task 4 in the ECO in Communications

122. d. Regulatory compliance

   Since the telecommunications industry is regulated, regulatory compliance is a key criterion.

   Portfolio Management Standard, pp. 66–67

   Task 4 in the ECO in Governance

123. c. A large number of concurrent programs and projects

   Resources tend to be limited in organizations, including funding for external resources. If a large number of concurrent programs and projects are under way at the same time, risks can arise if some of the same resources are required to support them or if there are dependencies among them that affect deliverables and benefits.

   Portfolio Management Standard, p. 120

   Task 1 in the ECO in Risk Management

124. d. Assist in scheduling adjustments

   The software also can help to provide data to resolve resource conflicts, and the dashboard report is useful to quickly communicate status. Resource utilization reports are especially helpful.

   Kerzner (2011), p. 295

   Portfolio Management Standard, p. 86

   Task 3 in the ECO in Performance

125. a.   Satisfying important information needs of stakeholders

Stakeholders at all levels require information as to why portfolio management is important and their own roles and responsibilities in the process.

Portfolio Management Standard, p. 105

Task 2 in the ECO in Communications

126. c.   The existing inventory of work should be validated against the updated strategy

The portfolio strategic plan aligns the portfolio with organizational strategy. To enable this strategic alignment, any pre-existing portfolios or inventory of work must be validated against strategy updates to ensure the plan is used to ensure organizational objectives are met.

Portfolio Management Standard, pp. 65–66

Task 1 in the ECO in Strategic Alignment

127. b.   Status in achieving benefits

As sponsors prepare proposals for inclusion in the portfolio, they emphasize tangible and intangible benefits of the component. Portfolio reporting is described as part of the portfolio performance plan.

Portfolio Management Standard, pp. 87, 111

Task 4 in the ECO in Governance

128. d.   Equity protection

The portfolio manager has a contingency amount, or equity protection, across the portfolio, which is used as a form of insurance if any component in the portfolio cannot fund its own contingency for risks. If a component has difficulty and lacks contingency funds, it is then funded from the equity protection fund. The entire value of the portfolio must be considered, and not the individual components, meaning if some of the programs or project fail, the portfolio is still a success if the overall portfolio provides benefits and business value.

Portfolio Management Standard, p. 120

Task 7 in the ECO in Risk Management

129. a. Included in the portfolio roadmap

The roadmap contains a list of components that provides a high-level strategic direction and information in a chronological fashion for portfolio execution.

Portfolio Management Standard, p. 39

Task 8 in the ECO in Strategic Alignment

130. c. Enable comparison among components

The purpose of a prioritization model is to use it to score and compare portfolio components to establish the portfolio and evaluate it.

Portfolio Management Standard, pp. 44–45

Task 6 in the ECO in Strategic Alignment

131. a. Employ a resource management process

To best support capacity and capability analysis, a resource management tool, technique, or process should be used to enable analysis.

Portfolio Management Standard, p. 90

Task 7 in the ECO in Performance

132. b. Business imperatives

This category fits this situation in that it covers the internal toolkit, IT compatibility, and upgrades; all of which are needed in this company to continue with its existing products as well as focusing on new products requiring technology upgrades.

Portfolio Management Standard, p. 68

Task 4 in the ECO in Strategic Alignment

133. a. Alcohol, tobacco, and sugar soft drink products should no longer be offered

The portfolio manager's risk concerns include risks that threaten the link between organizational strategy and the portfolio. In this situation, the company has added extensive health care services but has continued to provide products with known adverse health problems. There is now a dichotomy between these products and the health care services and is a risk at the portfolio level concerning overall organizational strategy.

Portfolio Management Standard, p. 122

Task 3 in the ECO in Risk Management

134. a. Work with your business partner in terms of portfolio management

It is appropriate in small organizations to have executives assume all or some of the portfolio governance responsibilities.

Portfolio Management Standard, p. 26

Task 1 in the ECO in Governance

135. a. Prepare a model of the current configuration and modify it to determine future capacity requirements

As the results of the model are analyzed, if the results indicate there is insufficient capacity to meet future requirements, it can be used to evaluate alternatives to find the best way to provide sufficient capacity.

Portfolio Management Standard, pp. 48, 74

Task 5 in the ECO in Strategic Alignment

136. c. Portfolio communications strategy

The portfolio communications strategy should be transparent and should mitigate any risks of lack of communications. Through a transparent approach one can see the priority of his or her work and how it relates to organizational strategy and objectives.

Portfolio Management Standard, p. 105

Task 2 in the ECO in Communications

137. d. State this requirement in the communication plan

This requirement is an example of a constraint, and it along with any other organization policies on communications should be in the portfolio communication plan.

Portfolio Management Standard, p. 113

Task 4 in the ECO in Communications

138. c. Provides guidance on stakeholder engagement

The portfolio management plan provides guidance in the extent of stakeholder engagement required to develop the portfolio risk management plan. Guidance also is provided concerning the governance approach, overall portfolio performance, and communications.

Portfolio Management Standard, p. 124

Task 2 in the ECO in Risk Management

139. c. Organizational process assets

Organizational process assets influence the portfolio's strategic plan and provide information to better understand the overall organizational strategy.

Portfolio Management Standard, p. 44

Task 4 in the ECO in Strategic Alignment

140. a. Identifies internal and external dependencies

One key benefit of the roadmap is its ability to show dependencies as decision makers then can determine the effect of adding a component or terminating one.

Portfolio Management Standard, p. 49

Task 8 in the ECO in Strategic Alignment

141. c. Portfolio management plan

This document is an input to the Manage Portfolio Risks process. It requires understanding of overall portfolio funding. It also contains the portfolio risk management plan, or this document is a subsidiary plan to it, which describes the time and budget for portfolio risk management.

Portfolio Management Standard, p. 131

Task 4 in the ECO in Risk Management

142. b. Revise the portfolio management plan

Among other things, this plan includes prioritization along with balancing the portfolio and managing dependencies.

Portfolio Management Standard, p. 39

Task 4 in the ECO in Governance

143. b. Included in finite capacity planning and reporting

Resource leveling is useful to demonstrate the effect if resources were not over allocated and to help make decisions on the best ways to prioritize work or to determine the types of resources to acquire.

Portfolio Management Standard, p. 90

Task 7 in the ECO in Performance

144. b.   Ensure the criteria and categories are aligned with the portfolio roadmap

The portfolio roadmap is used help identify components, categorize them, and determine scoring criteria from its existing information. This means as a minimum the criteria and categories require alignment with the roadmap.

Portfolio Management Standard, p. 66

Task 8 in the ECO in Strategic Alignment

145. b.   Portfolio algorithms

Portfolio process assets are an input to the Develop Portfolio Risk Management Plan process and contain specific portfolio information to aid in developing this plan. An example is the algorithms used in scoring components and prioritizing them.

Portfolio Management Standard, p. 124

Task 2 in the ECO in Risk Management

146. c.   Meet the organization's objectives

The communication's strategy ensures important stakeholder information needs are met. This information as it is collected is used to ensure the organization's objectives are met.

Portfolio Management Standard, p. 105

Task 1 in the ECO in Communications

147. c.   Assess gaps in meeting the Agency's strategic objectives

This is an initial list of the work under way, and it will require maintenance; however, with the list, it now is possible to determine what is being done and whether there are any gaps to fill to meet Agency objectives.

Portfolio Management Standard, pp. 43–44

Task 4 in the ECO in Strategic Alignment

148. b.   Enterprise environmental factor

In preparing or updating the performance management plan, enterprise environmental factors must be considered and require monitoring.

Portfolio Management Standard, p. 61

Task 4 in the ECO in Governance

149. a. Risk checklists

Updates to organizational process assets are an output of the Develop Risk Management Plan process. Often, organizations have risk checklists to make sure nothing that has occurred in the past has been overlooked as the risk management plan is prepared. New risk categories or sub-categories also may require updates.

Portfolio Management Standard, p. 129

Task 2 in the ECO in Risk Management

150. b. A, D, and C

These programs and projects have the greatest benefits associated with them and also are the ones that have the least risk.

Turner, Huemann, Anbari, and Bredillet, pp. 120–121

Portfolio Management Standard, p. 51

Task 7 in the ECO in Strategic Alignment

151. d. External resource providers

While they are a type of supplier, they focus only on providing resources to complement what is available or to perform a key function, such as resources for a Project Management Office or for IT.

Portfolio Management Standard, pp. 106–107

Task 1 in the ECO in Communications

152. c. A way to set up a common set of decision filters

Through the use of categories, a common set of decision filters and criteria can be used for evaluating, selecting, and prioritizing components.

Portfolio Management Standard, p. 64

Task 3 in the ECO in Strategic Alignment

153. d. The possible expected level of return for its level of risk

This approach is useful as a tool and technique in the Manage Portfolio Value process, and it is an analytical tool to analyze portfolios in light of resource constraints.

Portfolio Management Standard, p. 102

Task 6 in the ECO in Performance

154. c. Weighted ranking and scoring techniques

These techniques are used by the portfolio manager and the governance oversight group to assess risks in multiple portfolios in the organization and to the overall portfolio structure. They can be applied to any technical and management risk details as a tool and technique in the Develop Portfolio Risk Management Plan process. The high-level plans for risk management thus are defined by ranking and scoring,

Portfolio Management Standard, p. 125

Task 2 in the ECO in Risk Management

155. c. Compare the current portfolio mix with that with this change

A gap analysis is used to compare the current mix of components with the new strategy and the organization's 'to-be' vision.

Portfolio Management Standard, p. 54

Task 7 in the ECO in Strategic Alignment

156. b. Facilitates impact analysis

In change management impact analysis that may be part of the proposals for new components is used to then lead to reviews of the component and either its approval or disapproval.

Portfolio Management Standard, p. 63

Task 4 in the ECO in Governance

157. d. Impact the availability of work managed in the portfolio

Ideally, all work is managed by the portfolio, but in most organizations it is limited to programs and projects. Then when working in matrix or functional organizations, resources may be committed to both program/project work as well as operational work.

Portfolio Management Standard, pp. 92–93

Task 7 in the ECO in Performance

158. b. $50,000

By setting up a cumulative cost chart for the investment in the portfolio over a certain period, one can then see the level of investment that is required for certain probabilities of success.

Portfolio Management Standard, p. 103

Task 1 in the ECO in Risk Management

159. b.   Portfolio process assets

In conducting a strategic alignment analysis, its results then are part of the portfolio process assets and can be referenced later and as needed by the portfolio manager.

Portfolio Management Standard, pp. 54–55

Task 3 in the ECO in Strategic Alignment

160. b.   Provides information about interdependencies that may affect objectives

The roadmap is an input to the Plan Communication Management Plan process. Since it includes interdependencies among the components, it is useful to review as this information may affect communications objectives.

Portfolio Management Standard, p. 108

Task 2 in the ECO in Communications

161. b.   Decided to use a an outside facilitator when meetings are held

In the Provide Portfolio Oversight process, elicitation techniques, including facilitation, often are used to provide oversight.

Portfolio Management Standard, p. 83

Task 1 in the ECO in Governance

162. d.   Use a standardized format across components

Portfolio reports are an input to the Manage Portfolio Value process. While a variety of reports can be used, the objective is to ensure the reports support content and follow a standardized format across components.

Portfolio Management Standard, p. 98

Task 3 in the ECO in Performance

163. a.   Evaluate if the benefits of the portfolio are aligned with organizational strategy

Reports on benefits are useful as they explain whether there is utility to the organization and other beneficiaries. Therefore, these reports show alignment with strategy. If the components are not aligned with strategy, the portfolio process is not adding value, and changes are needed.

Portfolio Management Standard, p. 83

Task 1 in the ECO in Governance

164. a. Portfolio management plan as it shows all elements in it have communications requirements

    The communications management plan often is a subsidiary plan to the management plan or is part of it. Regardless, the management plan is the guiding artifact to consider, and it shows communications requirements of all elements in it such as risks that require communications to a variety of stakeholders.

    Portfolio Management Standard, p. 109

    Task 2 in the ECO in Communications

165. d. The overall value of the portfolio is affected adversely

    The roadmap is an input to the Manage Portfolio Value process. It can show a ripple effect when there are changes to the components, such as delivery delays or cancellations, thereby affecting portfolio value.

    Portfolio Management Standard, p. 98

    Task 8 in the ECO in Performance

166. b. A portfolio process asset

    As an output of the Provide Portfolio Oversight process, portfolio process assets may need updating, such as the decision register and open issues.

    Portfolio Management Standard, p. 84

    Task 2 in the ECO in Governance

167. d. Define the portfolio vision and objectives to align with organizational strategy

    The purpose of setting up a portfolio is to ensure all work in progress supports organizational strategy as resources are limited, and the components require justification and alignment.

    Portfolio Management Standard, pp. 43–44

    Task 1 in the ECO in Strategic Alignment

168. d. Portfolio process assets

    This is one of several portfolio process assets that should be considered as an input to the Plan Communication Management process to recognize roles and responsibilities.

    Portfolio Management Standard, p. 109

    Task 3 in the ECO in Communications

169. c. Are considered portfolio process assets

    Meetings of the Board tend to be scheduled on a regular basis; however, ad hoc meetings, typically triggered by external events, can be held. Portfolio process assets may include historical information, governance decisions, and open issues.

    Portfolio Management Standard, p. 83

    Task 1 in the ECO in Governance

170. d. Level of influence and level of interest

    This approach, set up in a quadrant, can show the types of stakeholders who want active communications or extensive information about the portfolio versus those that require only minimal information or wish to be kept informed.

    Portfolio Management Standard, p. 111

    Task 3 in the ECO in Communications

Practice Test 2

This practice test is designed to simulate PMI®'s 170-question PfMP® certification exam. You have four hours to answer all questions.

INSTRUCTIONS: Note the most suitable answer for each multiple-choice question in the appropriate space on the answer sheet.

1. An important but often overlooked item that should be part of the portfolio strategic plan is:

   a. Organizational structure and organizational areas
   b. Application of resources to components
   c. Portfolio roles and responsibilities
   d. A high-level timeline

2. While project management is becoming a standard way of working in many organizations, its success requires portfolio management to ensure every project in the portfolio supports organizational strategies. Portfolio management, however, requires a structure for effectiveness. In developing this structure, a best practice is to base it on:

   a. Availability of funding
   b. Specific categories
   c. Portfolio, sub-portfolios, programs, and projects based on organizational areas
   d. Specific portfolio management processes to be executed

3. You are the portfolio manager for your chicken feed company located in Europe. The company's strategic goal is to be the market leader and to support as many poultry plants as possible. The company is striving as well to acquire smaller competitors. The challenge is that with each acquisition there are risks involved including review of competitor products, their resources, and planned products. Your firm just acquired another company, which means all the portfolio documents must be updated. Also, the portfolio must be updated, and components reprioritized. You are working on the risk management plan. You are determining the effect of changing three factors in the portfolio so you are:

   a. Rebalancing the components
   b. Using trade-off analysis
   c. Using sensitivity analysis
   d. Estimating portfolio risk exposure

4. Assume your organization's strategy is undergoing a major change as executives realized the existing dog food products it had manufactured for years were no longer being purchased in the typical quantities, and it was losing market share to the competition. The executives are changing to a new line of products as well as entering the cat food market. Because of the new strategic direction, it is necessary to consider:

   a. Employee morale
   b. The need for an infusion in capital
   c. Requirements for training and developing new competencies and skills
   d. Key stakeholder expectations and communication requirements

5. Stakeholder analysis is critical in ensuring the stakeholders are part of the portfolio process and also to ensure their information needs are met. Not to be overlooked as one performs a stakeholder analysis is the:

   a. Relationships among them
   b. Frequency information is needed
   c. Communication vehicles to use
   d. Communications effectiveness

6. Assume you are performing a prioritization analysis to help provide advice to your Portfolio Review Board. To accomplish this analysis of five possible programs projects, and operational activities to pursue, your first step should be to:

    a. Define the portfolio roadmap

    b. Compare strategic objectives

    c. Identify dependencies between these five proposed components and others under way in the portfolio

    d. Perform a strategic assessment

7. Working in the regulated food industry in a large dairy cooperative, among your roles as the portfolio manager is to measure the portfolio performance and especially to do so to manage portfolio value. The most useful technique is:

    a. Earned value

    b. Variance analysis

    c. Trend analysis

    d. Benefits analysis

8. Your health information services and systems company recently won PMI's coveted PMO of the Year Award. The PMO has been in existence for eight years and has focused on continuous improvement as it conducts maturity assessments every two years. In the last maturity assessment, conducted a year before submitting for the award, the consultants pointed out that many portfolio management components were not aligned with organizational strategy. The PMO Director, who has portfolio management responsibility, realized that this meant:

    a. The company's workload had increased

    b. A defined business case was needed for each component

    c. Additional subject matter experts needed to be hired to augment the staff

    d. A portfolio performance plan was a necessary first step

9. Assume you manage your organization's day-to-day activities in the marketing department. Occasionally, though, your Department has a specific project; one example is its decision to undertake a business development maturity assessment. This situation shows you are then:

    a. A member of the Portfolio Selection Committee

    b. A project sponsor

    c. Interacting actively with the Portfolio Manager

    d. Managing relationships with other operational groups so this project attains needed resources

10. Assume you have categorized the various risks that could influence the portfolio. You want to determine the risk value to the overall portfolio. You want to compare the risks without representing the interface between them and present a graphical view for the members of your Portfolio Review Board, and you know new risks will arise as strategic objectives change. To provide these data, you prepare a(n):

    a. Risk histogram
    b. Influence analysis
    c. Heat map
    d. Impact assessment

11. Assume your organization had a slight change in strategy as it acquired a minor, small competitor with similar product and service offerings. This acquisition in essence adds resources to your company, but its products may have a few different features than those your medical device and health services company offers. As the portfolio manager, you performed a gap analysis to best manage strategic changes and to review the component mix in the portfolio. You then reviewed the portfolio roadmap and decided to update it because of changes to the:

    a. Management approach
    b. Priorities
    c. Components
    d. Risks

12. Recently, you were appointed as the portfolio manager for the county government. You were pleased you were selected for this opportunity. Previously in your work as a project manager and then as a program manager, you found a charter to be especially useful and plan to prepare one for your work in portfolio management. It should contain:

    a. The high-level timelines for portfolio delivery
    b. The prioritization criteria
    c. Key assumptions, constraints, and dependences
    d. A list of portfolio benefits

13. After realizing the organization's traditional methods of managing its programs and projects were too cumbersome and receiving complaints from customers that the deliverables did not meet their expectations or were late, your executive team decided to make some changes. They established a PMO and also had an external consultant review the existing procedures and simplify them. Now, staff members require training in them, and it seems as if with the simplified methods, some staff will be under-allocated. As the portfolio manager, you believe the executive team would find it useful for you to prepare:

    a. Financial investment reports
    b. Customer satisfaction surveys
    c. Resource histograms
    d. Resource profiles

14. Your manufacturing company focuses on a policy of 'zero defects'. However, for it to remain the market leader in its production of metal hip replacements, it has invested in a number of tests to ensure that once the product is delivered, defects are not present, leading to less rework and on time delivery. Since rework previously was the norm, and many claimed the previous product line led to injury, the company decided to focus on root-cause analysis as a way to identify problems and determine solutions. It used focus groups to involve a representative number of people from a variety of areas in the plant. Such an investment in quality:

    a. Demonstrates management commitment to the components in the portfolio
    b. Is viewed as a positive risk
    c. Is a way to reduce overall life cycle costs
    d. Involves everyone in the process

15. Recognizing the overload on the electrical power grid system in your country and the numerous outages that occur in metropolitan areas, you have been tasked by the Department of Energy to serve as its portfolio manager and to set up detailed identification, categorization, ranking, and scoring criteria. You found that an especially useful document to assist you was the:

    a. Department's strategic plan
    b. Portfolio strategic plan
    c. Portfolio charter
    d. Portfolio roadmap

16. Each time the Portfolio Review Board meets, components, both proposed and those in progress, are reviewed in terms of risk. The Board members are typically risk adverse especially since the company is the market leader in the package delivery market. However, during the recent holiday season, it was unable to keep up with the demand as there was tremendous on line ordering, and the result was bad publicity and extensive customer complaints. The Board is meeting on Friday to determine what actions to take to minimize these problems. The goal is to:

    a. Add new components and reallocate resources
    b. Minimize portfolio risks
    c. Analyze how to best optimize the portfolio
    d. Make decisions concerning existing components

17. Assume you are the sponsor for a new component in your government agency. The agency has a defined portfolio management process, and a Portfolio Oversight Group comprised of the agency executives and chaired by the Administrator. It meets regularly to determine whether existing components in the portfolio should be included and whether new components should be added. As the sponsor, you must follow the Agency's standard set of key descriptors as you present your proposal to the Portfolio Oversight Group. Such descriptors:

    a. Ensure all components are comparable
    b. Provide a categorized list for ongoing evaluation
    c. Address similar Agency needs
    d. Optimize the portfolio for value delivery

18. Assume you are the portfolio manager for one of the leading high technology companies in Silicon Valley, California. Change is constant as your company works in many areas to be the market leader in its products and services. The Portfolio Review Board meets weekly to approve new components, decide whether others should be terminated, and to rebalance the portfolio. One goal is to have breakthrough platforms that lead to greater value. A key to success for successful benefits realization from this changing portfolio is to:

    a. Have an easy-to-use PMIS
    b. Involve stakeholders
    c. Ensure the communications messages are consistent after each meeting
    d. Have an easy-to-use rebalancing method

19. Assume you are the portfolio manager for your research and development business portfolio. It is a critical role as this portfolio represents the breakthrough initiatives to be pursued by your automotive company. While you were not involved in creating the company's strategic plan, you do monitor changes that may affect the organization, and also you:

   a. Determine the portfolio algorithm to be used
   b. Determine the organization's strategic objectives
   c. Provide information on progress and results
   d. Select the decision support system to use

20. For many companies difficult market conditions can represent a window of opportunity in portfolio management. Long-term success may mean acquisitions as well as shredding or divestiture of some lines of business. Divestitures then can raise capital for acquisitions. Locating such windows of opportunity mean:

   a. Recognizing portfolio dynamics
   b. Focusing on interdependencies between high-priority portfolio components
   c. Enhancing potential improvements in the performance of portfolio components
   d. Focusing on generating new portfolio components

21. Assume your dairy cooperative is new to portfolio management. It decided to embrace it as an audit from the Food and Drug Administration showed that the cooperative was using inappropriate animal drug residues to test for milk safety. The audit also revealed the cooperative had not fully implemented mandatory Hazard Analysis and Critical Control Point procedures. Your executive team realized its focus was too heavily oriented toward existing products and being the first to market with new products, and as a result, it overlooked some components that were mandatory. Now you have put together portfolio management practices and a portfolio management plan, and the executive team is serving as the Portfolio Oversight Group. It is meeting on Friday to optimize the portfolio. At this meeting you should:

   a. Provide each member with a copy of the Portfolio Model
   b. Have the portfolio roadmap available
   c. Discuss the scoring approach you and your staff used to make recommendations as to new components to pursue and others to terminate
   d. Provide each member with an inventory of all the work in the portfolio

22. As you work as a portfolio manager, you have a number of activities to perform. However, you also recognize they:

    a. Reflect on the investments made or planned by the organization
    b. Identify and align organizational portfolios
    c. Measure value versus benefits
    d. Assist in managing risks and communications

23. You have four possible components to recommend as you work to optimize the portfolio. There are limited resources in your company, which prides itself on time to market as an attribute that distinguishes it from its competitors in the automobile parts field. Each component contributes to the organization's strategy. Component A will eliminate features if there is a trade-off situation. Component B will delay its schedule if necessary. Component C has a flexible structure with a focus on innovative features at a minimum cost. Component D plans to focus on technical, cost, and schedule as its metrics to report regularly. You recommend:

    a. Component A
    b. Component B
    c. Component C
    d. Component D

24. Recently, your company, a low price but quality department and food store, had credit card theft, and many customers across the country were affected adversely. It had not been focusing on brand recognition as an organizational value area, and now the number of customer complaints is at an all-time high. Metrics are necessary in the area of customer relationship management, and new components to repair the quality brand image for customers are required to be added to the portfolio. This situation shows:

    a. These new metrics will not be quantifiable and therefore are hard to fit the "SMART" criteria used for portfolio management
    b. These metrics will be difficult to collect and achieve positive gains in a short time frame
    c. The portfolio management office is ideal to develop and collect the new metrics
    d. The damage to repair the brand image is such that it may be impossible to completely recover

25. Desiring a federal government career representing security and safety in terms of downsizing and budget cuts that are so prevalent, assume you left your position as a portfolio manager at a leading global training firm and joined the Census Bureau as its portfolio manager. In addition to the country's national census every four years on the number of people in the country, the Bureau conducts other types of censuses periodically. Most of the employees have been in this Bureau for their entire careers as it is considered a 'safe' place to work, and a risk adverse culture prevails. However, current executives are concerned that other initiatives should be pursued so the organization remains independent. This approach means that:

   a. New initiatives should be ones that are considered to be in the high risk/high benefit category
   b. Strategic objectives must be optimized
   c. An orientation on risk management as both threats and opportunities should be available to employees
   d. The new approach represents a major culture change, and a targeted communications strategy

26. Working to prepare the portfolio risk management plan, you asked those stakeholders who have been active proponents of portfolio management since the CEO announced it was being implemented to work with you. The stakeholders represent a cross section of the company's eight business units. Since the portfolio management plan has been prepared, approved, and distributed, your team is reviewing it as it contains:

   a. Stakeholder risk tolerances
   b. Organizational risk tolerance
   c. Portfolio resources
   d. Portfolio components

27. Each time the portfolio is updated with changes to authorized components, then documentation is available so work on them can begin. As well, a best practice to follow as the portfolio manager is to:

   a. Provide information to those in portfolio component support functions
   b. Update the inventory of existing work under way
   c. Update the roadmap
   d. Review the organizational process assets to see if updates are required

28. Your organization uses a funding process for components in the portfolio in which funding is divided between components based on contractual milestones for the components. This approach has been effective as some components require a lot of funding immediately to meet a contractual milestone, while others may be in the design phase for some time and not need as much funding immediately. This is an example of:

    a. Information in a report used to authorize the portfolio
    b. A funding reconciliation method so that each component in the portfolio receives funds when they are required
    c. An approach to monitor and track portfolio components based on key milestones
    d. A method to effectively use the roadmap to project funding requirements

29. Assume you are preparing your portfolio communications management plan. You want to structure it in a way that is the most meaningful for your stakeholders. Therefore, your first section should:

    a. State the recipients of the plan
    b. Describe goals and objectives
    c. Link the plan to the portfolio strategic plan
    d. Present the results of your stakeholder analysis

30. Assume you have been asked to create some portfolio scenarios to aid your Portfolio Review Board at its upcoming meeting as the company is facing a 20% budget cut. You prepare an expanded version of the roadmap, which is an example of:

    a. An interdependency analysis
    b. A prioritization analysis
    c. A sensitivity analysis
    d. An affinity diagram

31. Your company specializes in accessories for smart phones. Recognizing that phone companies strive to add new features to phones to stay ahead of the competition, your company has a strategic goal to ensure its accessories can be used on future products without change. To do so as the portfolio manager, you focus on a number of marketing factors as you recognize that changing one of these factors may affect the entire portfolio or the portfolio's strategy. To do so, you use:

    a. Trend analysis
    b. Time-to-market variability
    c. Market-payoff variability
    d. Market requirement variability analysis

32. You are the portfolio manager for the Department of Education and report to the Department's Secretary. She chairs a Portfolio Review Board, which meets quarterly, but you provide monthly reports to her and the other Board members on the portfolio's performance, and often ad hoc meetings of the Board are held after they read these reports. The purpose is to provide feedback that then:

    a. Program and project managers can use to ensure their work remains in strategic alignment
    b. May cause the other Board members representing the different units in the Department to rebalance their portfolios
    c. Lead to the potential change of the Department's strategic direction
    d. Provide awareness of potential internal and external impacts to Departmental goals

33. Two years ago, your University set up an Enterprise Program Management Office (EPMO) for its programs and projects. It now has repeatable processes in place, and the programs and projects are able to be completed within the allotted budget. However, the Dean of Graduate Studies recently has convinced the Chancellor the University needs to offer two additional doctoral degrees and a campus should be established in China. The Chancellor then met with the EPMO Director and decided to set up a separate portfolio office. You applied for this position and were selected. You are preparing your portfolio strategic plan. You know it is a best practice to include in it the:

    a. Organizational communication strategy
    b. Expected timelines
    c. Methods to optimize the portfolio
    d. Governance model

34. Assume a new Chief Financial Officer just joined your company. You are the portfolio manager, and he will be a member of the company's Portfolio Review Board. He will attend his first Board meeting in one week and asks you about the approach that is used to manage the portfolio to meet the corporate strategies. Your next step is to:

    a. Ask for a meeting to explain how portfolio management is practiced in the company
    b. Provide a copy of the portfolio management plan
    c. Provide a copy of the portfolio strategic plan
    d. Provide a copy of the Board's decision register

35. Change is constant in your telecommunications company because of consumer demand for new features in your products, competition, and regulatory requirements from the Federal Communications Commission. As the portfolio manager, you are the lead in managing the strategic changes that do occur. One item to review or update in this process is:

    a. The composition of the Portfolio Review Board
    b. Reports and other data provided to the Board for their consideration as they revise the mix of components
    c. The original portfolio charter
    d. The prioritization analysis methodology

36. Setting up and implementing portfolio management is difficult regardless of the size of the organization. Further, many organizations have multiple portfolios at different levels. An underlying principle of portfolio management is to focus on strategic initiatives. In doing so in planning, a best practice is to:

    a. Involve as many stakeholders as possible
    b. Follow a strategic mandate
    c. Balance conflicting demands relative to programs and projects
    d. Manage necessary changes

37. In determining the list of portfolio components that best aligns with organizational strategy, one approach is to use capacity analysis. This approach is especially useful when there are internal constraints. Assume your organization has decided to move into a management-by-projects environment and trained its engineering, manufacturing, and IT staff in a variety of project management techniques including how to develop a WBS, project plan, quality plan, and risk management plan. You have been asked as a consultant to assess the organization's ability to pursue projects that cross functional disciplines. You recommend:

    a. Use of external resources
    b. Mentoring and job shadowing
    c. Determining constraints from capabilities
    d. Use of the trained internal resources

38. Assume you are working for a University that offers a masters degree in project management. While it was one of the first Universities in your country to offer this degree, now about 75 other Universities are offering degrees or certificate programs. Your University requires that all courses be taken on campus; it does not have a distance learning program. As a result, enrollments have decreased even though it is a quality program. The decreasing enrollments now are viewed as a negative risk to continue the program at the University. You are assigned as the risk owner and are analyzing this risk through probability and impact. You should:

    a. Discuss with the Portfolio Review Board that unless an on-line component can be added, enrollments will continue to decrease

    b. Use the probability and impact ratings as stated in the portfolio risk management plan

    c. Note that if tuition costs are decreased, enrollments will stabilize

    d. Complement the probability/impact analysis with sensitivity analysis before making recommendations

39. Assume your Portfolio Review Board completed its analysis of the existing components in the portfolio and decided to terminate a new product for a new stent to assist heart patients. At the time it was approved, its features far surpassed any on the market, and it seemed since safety and quality are the key priorities, regulatory approval time would be minimal. However the Center for Devices in the Food and Drug Administration found a major flaw in your company's product, and now a competitor has a comparable product on the market. As the portfolio manager, it is your responsibility now to notify the program manager and team that the program is canceled, and most will lose their jobs as a result. In this situation, your best course of action is to:

    a. Make sure the people involved know the decision was one the organization's executives made as members of the Portfolio Review Board

    b. Let the team know the termination is due to the competition being first to market

    c. Use the human resources person on the Portfolio Review Board and ask him to navigate the negative impact on the people involved

    d. Personally offer assistance to anyone on the team in locating work and provide references stating why the project was terminated

40. Assume you want to assess the efficiency in terms of the actual costs of the various resources working on components in your portfolio. You have set up a report to show the time and cost for people working on components since they tend to work on multiple programs and projects. One approach to use is to prepare:

    a. A cause-and-effect diagram
    b. A funnel chart
    c. A productivity index
    d. A key performance indicator

41. Your organization is reviewing its strategic goals and objectives as well as the prioritization criteria used for portfolio decisions. The Chair of the Portfolio Review Board now is requesting that each existing component and proposed components be evaluated to determine how they support customer satisfaction, increasing revenue, attracting new customers, and the degree of strategic alignment. This approach means:

    a. Components must support one of these four areas
    b. Components will be terminated if they do not support one of these areas
    c. Components may support more than one of these areas
    d. The existing portfolio scoring model will not be needed

42. Assume you are preparing reports on the status of the portfolio. After conducting your communications analysis, you learned your key stakeholders wanted a graphical report that showed a variety of information on a single page. They were interested in schedule status, issues and risks, and financial information so you decided to:

    a. Set up a traffic light report in these areas
    b. Use a quadrant report with these items representing overall success
    c. Use a dashboard with summary information in these areas
    d. Determine KPIs for each area and then prepare a quadrant report

43. Assume you work as the portfolio manager for a company specializing in food additives. A component is to develop an additive to reduce obesity and still enable people to eat as much ice cream and chocolate as they want. However, a competitor just released a similar additive. Your Portfolio Review Board is holding an ad hoc meeting to determine next steps. You had identified this possibility as a potential risk. You worked with the risk owner and determined the size of the investment in this component to date. You then decided the best approach was to accept this risk in an active way, which the Portfolio Review Board supported. This means:

   a. You will go ahead as planned with this component
   b. You will get approval to add more resources to complete the product development as quickly as possible
   c. You will decide to terminate further work given the low level of investment and sunk costs to date
   d. You will use your contingency reserve and work aggressively to complete the component

44. Your company is using mergers and acquisitions as a means of entering new markets in the food additive area as well as eliminating competition from existing products. As the portfolio manager, it seems that you are consumed by change and are constantly rebalancing the portfolio. As you do so, you need to maximize portfolio return according to the company's:

   a. Strategic goals
   b. Available resources
   c. Available assets or technology
   d. Desired risk profile

45. Different organizations have different definitions for vision, mission, and values. However, portfolio, program, and project management are the way an organization best can manage the capability to deliver value, which then is realized through:

   a. Customer relationship management
   b. End user satisfaction
   c. Day-to-day processes
   d. Benefit sustainment

46. Assume you are working to optimize your portfolio, and your management is interested in your using quantitative analysis as you do so. You are considering four possible components but only can recommend one given resource constraints Each one has benefits supporting the company's strategic goals and objectives, but you are to recommend the one with the shortest payback period. Project A is estimated to cost $100,000 to implement with annual net cash inflows of $25,000. Program A is estimated to cost $75,000 with inflows of $20,000. Program B is estimated to cost $225,000 with inflows of $80,000. Project B is estimated to cost $275,000 with inflows of $90,000. You have net present value (NPV) data available as follows:

Project A NPV at	Program A NPV at	Program B NPV at	Project B NPV at
5% = $2,399	5% = $2,105	5% = $6,400	5% = $4,065
10% = $3,112	10% = $1,254	10% = $3,275	10% = $1,852
15% = $1,402	15% = $1,001	15% = $1,679	15% = $925

Based on these data, you recommend:

a. Project A
b. Program A
c. Program B
d. Project B

47. Although there is a talent gap in the program and project management fields, in your government agency because of the economic downturn, many people lost their jobs, and the challenge is to do 'more with less' to justify the agency's existence. As the portfolio manager, you have an inventory of the work that is under way and has been authorized. You also have documented acceptable ranges for optimal resource capacity utilization in your:

a. Resource management plan
b. Risk management plan
c. Performance management plan
d. Portfolio value plan

48. Assume the Portfolio Review Board met and made some recommendations for changes to five portfolio components as it passively accepted changes to three components and actively accepted changes to two others. As the portfolio manager, you need to:

a. Update changes in the portfolio component lists
b. Update the portfolio issues register
c. Reassess overall portfolio risks
d. Reassess overall organizational risks

49. Your railroad has been using a formal portfolio management process for more than 10 years; it also as an Enterprise Program Management Office for its programs and projects. One way the Portfolio Management Office can best work with and support the Enterprise Program Management Office is by:

    a. Using the Portfolio Office to negotiate for resources to support components
    b. Providing frameworks and methodologies to support portfolio, program, and project management in the railroad
    c. Forecasting resource supply and demand
    d. Assisting in risk identification and risk strategy development and communicating all risks and issues to all stakeholders

50. Assume your organization set up a pilot project in one business unit to see the benefits from a defined portfolio process during the past year. The pilot uncovered several projects that were under way that did not contribute to the organization's current strategies but had not been terminated. It also revealed gaps as some key objectives were not supported. The results of using portfolio management were positive, and as the portfolio manager for this business unit, you were selected as the portfolio manager for the company. You explained a key first step was to:

    a. Prepare a portfolio charter
    b. Set up a portfolio roadmap
    c. Set up a governance structure at the highest level chaired by the CEO
    d. Document communications required for successful implementation

51. Assume you are managing a portfolio in a business unit in your telecommunications company. You have set up the process to follow to authorize components to be in the portfolio and have an active Portfolio Review Board. One Board member, who recently joined the company, used dashboard reporting on portfolio status at his previous company and has asked you to prepare dashboard reports every two weeks. This means:

    a. The purpose is to chart progress
    b. In addition to visual graphs, you will provide comments
    c. Your focus is on events
    d. The reports are periodic snapshots

52. In your construction company, assume the Construction Industry takes each company in various markets and rates it from best in class to the worst in terms of overall customer satisfaction, quality, and health and safety. In the past three years, your firm's rating has been decreasing, and this year it was last on the list. Your executives now realize drastic action is required and is reexamining its portfolio. They have approved business cases for new programs and projects to reverse this trend. To do so, senior managers are leading these new components, a PMIS has been established, and you are leading a Portfolio Management Office. You prepared a portfolio strategic plan and an innovative communications plan. In four years, the firm moved to number two on this list. You attribute success to:

    a. Strong executive support
    b. Development of a robust yet customized portfolio management process
    c. Stakeholder engagement from the start
    d. A structured portfolio performance plan

53. Assume you work actively since you are the portfolio manager with your stakeholders to provide a framework that is best to operationalize the strategic goals and objectives. However, your oil and gas company recently acquired a major competitor, and it now is the largest in your country. Since this merger, the company has completely changed its strategy. You realize detailed stakeholder analysis is critical because:

    a. The merger means the composition of the Review Board will change
    b. Continuity and alignment of expectations are critical to success
    c. The mix of components in the portfolio will change
    d. You need stakeholder support for the portfolio function to continue

54. A PMIS can be a simple spreadsheet or a sophisticated software system. Since your organization has a mature portfolio management process in use, assume it has an enterprise software system to support it. This system is used by component managers, the portfolio manager and staff, the portfolio management office, and stakeholders at all levels, including executives. Such an advanced system is helpful in that:

    a. It links to the project management information system
    b. It enables resource optimization
    c. It has measures and templates for reporting
    d. It links to the staff members' knowledge, skills, and competency profiles

55. Assume you live in Canada where hurricanes typically are extremely rare except for remnants of those from the U.S. and Caribbean off Newfoundland. You are the portfolio manager for Canada's Weather Service. You have been tracking the number of hurricanes that have affected the country, and you found the trend is increasing as is the intensity of the storms. Some have affected Nova Scotia and Prince Edward Island. When the hurricanes strike, the Weather Service is ill equipped to provide sufficient warning for them as is done in the U.S. You feel that the Weather Service requires new components to be in the portfolio, and the Portfolio Oversight group agrees. This situation shows:

    a. The need for reports to address trends and show variance analysis
    b. The changing nature indicative of all portfolios
    c. The need to update the strategic goals and the corresponding portfolio strategic plan
    d. An emphasis on risk management may lead to new components

56. Assume you are the portfolio manager for your transit company, the largest in your country. You planned to expand the commuter train lines to serve more commuters and to set more lines up with higher rates of speed that were non-stop trains. Other new components involved expanding the bus service and making the buses more energy efficient. However, a major commuter train derailment caused many fatalities, and as a result, all new components are on hold pending safety inspections at the Federal, State, and City levels by mass transit officials. This means:

    a. Compliance programs now have the highest priority in your portfolio model
    b. The strategy for the transit company requires change
    c. You need approval to update the portfolio strategic plan
    d. The next step is to inform sponsors of the proposed components that they are deferred until further notice

57. Assume you are the leader of the portfolio planning team, and your team is working with the Portfolio Review Board to prepare templates and examples for performance measures and performance targets. Once these are in place the next step is to:

    a. Document them and provide them to component managers
    b. Establish a focus group for validation that these metrics can be collected easily
    c. Prepare an orientation session for interested stakeholders on these measures and targets
    d. Have the Portfolio Review Board obtain approval by the executive team

58. You are setting up a portfolio management process in your baby food company as the CEO wants to expand and also enter the pet foods business where it lacks any presence. As you work to set up portfolio management, you are defining how the company's assets and resources will be planned in the portfolio based on the corporate environment, which means you are:

    a. Preparing the governance model
    b. Using organizational process assets
    c. Following environmental factors
    d. Establishing portfolio assets

59. Although you used communication requirement analysis as you prepared your communications management plan, you plan to again use it to determine effective ways to manage portfolio information. One reason why it will be helpful is that it:

    a. Can determine the most effective ways to communicate portfolio status
    b. Can provide advanced warning to decision makers regarding market trends affecting the portfolio
    c. Can build forecasts based on trends in the data
    d. Can communicate multiple messages on portfolio status

60. Assume you are the portfolio manager of a leading department store that specializes in high-end clothing, furnishing, and other items. It is exclusive, and many cities in your country want a store. The existing stores recently added a bistro, and each one is recommended highly. However, last month your store was the victim of cyber security fraud across the country. You now need to actively restore the image of the store to continue to attract wealthy shoppers. You have a series of components to recommend, but after conducting a supply and demand analysis, you find that to sustain strategic alignment, you must recommend to the Portfolio Governance Board that it:

    a. Postpone some planned openings of three stores
    b. Put existing components on hold to reallocate resources to restore brand image
    c. Update existing portfolio resource allocations and schedules
    d. Provide new funding to the components to restore the store's image while continuing with the existing portfolio mix

61. Assume you are the portfolio manager for the leading toothpaste company. As a major sporting event soon will be under way in which counties from throughout the world will be participating, your company learned today that some feel toothpaste may be used to conceal explosives. You want to ensure your products are completely safe, and no tampering is possible so you decide to add a component and announce it through a press release. This is an example of a(n):

    a. Structural risk
    b. Execution risk
    c. Unknown unknown
    d. Need for a management reserve

62. Different stakeholders require involvement in governance activities at different times in portfolio management. In your new camel milk portfolio, a new venture for the dairy cooperative, phase-gate reviews should include representatives from:

    a. Engineering
    b. Manufacturing
    c. Research and development
    d. Legal

63. Your portfolio stakeholders review a number of reports each month to monitor the overall value of the value of the components to the business. One report of interest shows the assigned resources to components and then shows how much work is completed and how much is remaining with work tracked in person days. Such a report is a:

    a. Resource prioritization report
    b. Funnel diagram
    c. Histogram
    d. Burn-down chart

64. You are working to ensure you identify how often the various stakeholder groups require information about the status of the portfolio and how to best provide it to them. You also decide to conduct a review of your analysis in order to:

    a. Ensure redundant information is not provided
    b. Ensure consistent messages are delivered regardless of the frequency
    c. Determine how the information will be stored and retrieved
    d. Ensure all portfolio information is provided

65. Assume you are the portfolio manager for your natural gas distribution system company. It is considering acquiring a transmission company in the same area of the country, and this acquisition, once approved by regulatory authorities, will broaden the portfolio enormously. You have organized the portfolio for your company by categories, and this approach is beneficial as it:

    a. Enables appropriate balancing
    b. Provides a way to use software easily in prioritization
    c. Identifies those components in greatest need of key subject matter experts
    d. Ensures components in each category have a common goal

66. As the portfolio manager, you and your team collect a variety of tangible and intangible metrics to assess performance of the components in the portfolio. Each component manager submits reports to the sponsor, who consolidates them and sends them to you. You and your team then prepare a report for the Portfolio Review Board. As portfolio management has evolved in your new product development company in the past nine years, a metric now of interest is:

    a. Internal rate of return
    b. Net present value
    c. Sustainability
    d. Cycle time reduction

67. Having worked in the portfolio management field for more than 10 years, you realize having an up-to-date roadmap is extremely useful as it promotes portfolio management and shows the 'to-be' state in a way that is visually appealing. The dependencies and integration of the components in the roadmap also influence:

    a. The portfolio organizational structure
    b. The number of sub-portfolios in the organization
    c. The portfolio processes to follow
    d. The portfolio management plan

68. Previously, you were the portfolio manager for a consulting company, but given the downturn in the economy and its decreased revenues, you applied for and were offered the position of portfolio manager for a major pipeline company. It has not implemented portfolio management before, and you know a strategic plan is useful. You are finding its preparation to be more difficult than when you did so at the consulting company because:

    a. It is a completely new function
    b. There are external environmental factors to consider
    c. People require training in why portfolio management is needed
    d. The company's growth and ROI are strong

69. Assume you are working to prepare your portfolio communications plan. You have started the process of preparing your stakeholder analysis. You found many stakeholders are unsure as to how their work fits into the overall portfolio and want information about it. This means:

    a. You should set up a dashboard to portray the overall priorities in the portfolio
    b. A forecast of the portfolio's direction would be helpful
    c. The roadmap is useful to show the portfolio's structure
    d. An assessment of the value of the portfolio should be part of the plan

70. The dynamic nature of portfolio management makes it difficult to implement. If the organization has defined processes to follow and an engaged Portfolio Review Board that meets regularly, components will be added to the portfolio, and others will be terminated or deferred. When components are removed from the portfolio, the organization then:

    a. Revises its component categories
    b. Reallocates resources
    c. Revises the portfolio performance plan
    d. Revises the portfolio strategic plan

71. It is important to have a defined process in place to determine the priority of each component in the program for many reasons, one of which is to:

    a. Guide talent development
    b. Minimize risks
    c. Understand strategic objectives
    d. Document assumptions and constraints

72. Assume you have completed your risk probability and impact assessment. Since you work in a non-regulatory environment and also one in which the potential effect of legislation is low on your portfolio's components, these types of risks probably do not require additional work. However, a best practice is to:

    a. Continue to monitor them if the strategy changes
    b. Conduct some scenario analyses to determine whether or not further analysis is required
    c. Realize further investment is not needed on these risks
    d. Maintain them on a watch list

73. It is estimated that billions of dollars are lost annually because of software projects that cannot be completed or on other software projects that are completed but later are found to have effects. Often the problem is the lack of capability of the assigned resources to do the work, and defects are identified at the time the software is to be delivered. Because of the costs involved with resource misallocations, your company decided it would develop an economics-based model it then could sell to others to best allocate resources to software projects. Such an approach:

    a. Will not be helpful unless agile development will be used
    b. Can support resource capacity analysis
    c. Can model scenarios to use resources based on priorities
    d. Will be useful in dependency analysis

74. As a portfolio manager in an organization that is in the process of formal implementation of portfolio management, you reviewed a number of prioritization models and software packages that could be used. You then decided on a scorecard approach focusing on alignment to strategy, ROI, risk, and dependencies. This approach:

    a. Is documented in the portfolio strategic plan
    b. Is documented in the portfolio performance plan
    c. Receives stakeholder buy in from the communications plan
    d. Is part of the governance model

75. Your company has embraced portfolio management. Each business unit has a Portfolio Review Board that meets monthly for detailed reviews and prioritization, and the company's executives meet quarterly to prioritize at the corporate level. There are numerous artifacts to maintain, especially these governance decisions. As the corporate portfolio manager, you decided to:

    a. Establish a portfolio management information system
    b. Establish a portfolio management portal
    c. Maintain a governance decision register
    d. Establish a portfolio knowledge management system

76. Assume you are considering a proposed project to be part of your portfolio. But, when the Portfolio Review Board met, it realized it could not consider it as overhead because of its size, and it was not time to treat it as a capital investment because of financial constraints in your oil company. It also is considered to have many risks associated with it. You decide the best approach is to:

    a. Set it up to treat the project as an option
    b. Inform the project's sponsor to resubmit it at a later time
    c. Prepare a NPV analysis
    d. Use probability and decision-tree analyses

77. Assume you have decided to use interviews and questionnaires to help identify risks on an ongoing basis and to maintain a risk register. You plan to have a report available on the top ten risks that might affect the portfolio, either positively or negatively, and assess the probability and impact of the ten on this list. The list will change, and you then will update members of the Portfolio Review Board and other interested stakeholders. In your portfolio risk plan, you should:

   a. Document this approach in the methodology section
   b. Use subject matter experts to validate your approach before you prepare your risk plan
   c. Select risks to be in the top ten with at least one from your five risk categories
   d. Determine which risks will arise from various situations as part of your approach

78. Your goal as the portfolio manager is to optimize and balance the portfolio for greatest performance and also overall value delivery. This means you need to:

   a. Follow the portfolio strategic plan
   b. Communicate effectively with stakeholders at all levels
   c. Set up appropriate criteria to rank components
   d. Create the portfolio mix with the greatest potential

79. It is easy to focus on the benefits programs will deliver to the organization, and the deliverables projects will produce. Many organizations do not have a clear understand of all of the programs and projects under way, and many people do not want to disclose some 'pet' projects that are under way as they believe they are breakthrough initiatives for the company. Your organization lacks such an inventory and requires one as the executive team mandated that a portfolio management process be followed. The executives plan to meet monthly to determine whether or not new components should be added, and others deferred or terminated. The overall objective is to ensure:

   a. Components are focused on alignment to strategic objectives
   b. The portfolio strategy focuses on preventing poor return on investment in the components that are pursued
   c. Component benefits are emphasized along with deliverables
   d. The emphasis continues on the triple constraint as new components are pursued

80. As you work to manage supply and demand of resources in the portfolio, you need to determine whether there are any specific guidelines to follow especially if there are delays because of constraints on resources. You have found resources are often overbooked, leading to delays in completing your IT projects on time, and the company's reputation is suffering. These guidelines are contained in the:

    a. Organizational process assets
    b. Portfolio process assets
    c. Portfolio management plan
    d. Portfolio performance management plan

81. Your pharmaceutical company is noted for its best practices in project management. It also has embraced program management for greater overall benefits to its stakeholders and has had a defined portfolio management process in place for five years. As the company wants to continuously improve, recently it used an external consultant to perform an *OPM3* assessment. The company scored 80% in terms of best practices in place in portfolio management. The consultant was especially impressed with its portfolio management information system because it:

    a. Also linked to a knowledge management system
    b. Had sophisticated security procedures regarding access rights to key data
    c. Linked to the program and project management systems as an integrated system
    d. Included real-time dashboards

82. Your low end department store has entered the grocery business in many of its locations. It led to an increase in market share and a competitive advantage. Strategic organizational goals are to expand add stores to handle groceries, expand to other states, attract upscale designers to add their products, maintain its brand recognition of a clean and safe environment for shopping with friendly staff, and also to add bistros. However two weeks ago, there was a major security breach affecting customers in the stores throughout your country, and millions of people ended up with security fraud. On Friday, the Portfolio Review Board will review proposals for new components as well as progress with the existing components. But, resources already are stretched too thin, meaning some components will need to be terminated, or only one new component can be added. The Board should elect to:

   a. Add a component to restore brand image
   b. Terminate components in five states in your country to add groceries and add components for bistros to those stores now offering groceries
   c. Add a component to pilot test enhancing the company's image by attracting upscale designer clothes
   d. Add a component to redesign existing stores so their appearance is further enhanced

83. The portfolio risk management plan is used in many ways. Among other things, it updates:

   a. Portfolio funding
   b. Risk checklists
   c. Criteria to see if risks are identified consistently with the organization's risk strategy
   d. The various identified risk categories and any sub-categories

84. Assume your dairy cooperative, which spans 15 states in your country, has been focusing on improved project management and has an Enterprise Project Management Office. It is now moving into portfolio management and since you were a senior staff member in the PMO, you have been asked by the CEO to lead the portfolio management process planning and implementation. You are a team of one person at this time, and the executive team has asked you to first provide program and project information to them. To facilitate this request, you decide to:

   a. Send out a survey to the program and project managers
   b. Ask the PMO to gather this information for you
   c. Interview the program and project managers and use the interviews to explain why the cooperative is moving into portfolio management
   d. Ask each member of the executive team to provide a point of contact for you in his or her business unit or operations function and gather the data through this person

85. For purposes of continuous improvement, each year you review the criteria you are using to select and produce a categorized list of components in priority order. Your goal is to ensure the criteria you use are quantifiable, and in your web system development company, it is especially important to consider:

    a. Internal and external risks
    b. Technology capabilities and capacities
    c. Internal and external dependencies
    d. Legal and regulatory compliance

86. Assume your organization is undergoing a major transformational change. Some people have said it is similar in scope to the mid 1990s, when IBM changed from a focus on hardware to one focused on providing services and being project based. You are the portfolio manager and are leading this major transformation initiative. Many key stakeholders do not believe the people in the organization can accept the change and will resist it tremendously. You decide to:

    a. Hold a meeting through Web Ex to explain why the change is necessary
    b. Send out a message to every member of the organization so there is consistency
    c. Determine whether there are any needs that are required to best implement the change
    d. Meet with the key stakeholders and then set up some focus groups to gauge attitudes about the change

87. Your Portfolio Management Office prepares reports on the status of the components in the portfolio. These reports are helpful in terms of monitoring the portfolio especially in terms of supply and demand for resources. One report that often is overlooked is the:

    a. Capability report
    b. Benefit realization report
    c. Financial report
    d. Productivity report

88. Assume you have a governance model as part of your portfolio management plan. Over time, as portfolio management has matured in your organization and has been embraced by stakeholders at various levels, it is now a routine way of working. One reason it is effective is how portfolio governance is defined in that it prescribes:

    a. Ways to respond best to manage stakeholder expectations
    b. The overall strategic goals of the portfolio
    c. Internal resource allocation
    d. Risk response planning

89. Assume you are establishing the type and frequency of reports to be provided to various stakeholder groups as part of portfolio planning. You want to make sure the program and project managers, their team members, and any subject matter experts are not overlooked in the process. You decided they would have a specific interest in weekly information on:

    a. Resource allocation
    b. Prioritization
    c. Risks and issues involving the portfolio
    d. Governance meetings and decisions

90. Risks seem to be occurring that were not considered as part of your cereal company's portfolio. Many products have been recalled, which never happened in the past. You are working to manage risks to the portfolio and find useful information to help you in the:

    a. Risk breakdown structure
    b. Risk tolerance profiles
    c. Portfolio performance reports
    d. Portfolio investment choices

91. Assume you are the portfolio manager for your ball point pen company. The executives held their monthly meeting to review and optimize the portfolio and made a decision to add a new product line that requires new technology. These pens will be expensive, a pleasure to use, and will last forever. Another product line of pens in which the logo of corporations purchasing the pens, a product for which your company is well known, was terminated. Portfolio reports require updates especially in terms of:

    a. Internal and external environmental factors
    b. Organizational process assets
    c. Risk ratings
    d. Organizational areas

92. The portfolio charter includes many significant items. One that may be overlooked is:

    a. Issues
    b. Mission
    c. Stakeholder expectations
    d. Communication methods

93. Your health care company has been characterized by many as a tribal culture, as people work in various functional departments based on their scientific or medical background. However, you have a new CEO who previously worked in the defense industry and decided portfolio management was needed in your company for effective resource allocation. You were appointed as the Portfolio Manager and now have one person to help you. Your first step is to:

    a. Prepare a one-page vision statement for the CEO to issue to the organization as to why portfolio management is necessary
    b. Assess the current state of portfolio management
    c. Prepare a strategic plan for portfolio management
    d. Set up a Portfolio Review Board

94. Assume that you have been managing a portfolio in your publishing company for books on project management. Sales are not as high in the past two years even though you have e-books and PDF versions available and an aggressive marketing campaign. You realize there are so many project management books available from different publishers or from self publishing that this portfolio requires changes to continue to be viable. You must report the progress of this portfolio in two weeks to significant stakeholders in your organization and realize the best approach is to:

    a. Include an assessment of the probability of this portfolio to achieve its benefits
    b. Submit a proposal for additions to it in the areas of portfolio and program management books
    c. Recognize resources are limited and recommend reassigning some key SMEs to other portfolios to better handle demand
    d. Continue to ensure your own work on this portfolio is done by applying needed processes so your work is done effectively and efficiently

95. Recognizing the importance of an effective portfolio management process that people within the organization will embrace at all levels, you are leading this initiative at your University, which already has a Program Management Office. You decided to prepare a roadmap to best implement the structure and processes, but before doing so, you first worked with key stakeholders in order to:

    a. Define their expectations for portfolio management
    b. Determine the mission
    c. Prepare a vision
    d. Prepare a communications strategy to avoid mis-communication

96. Recognizing the need for specialized resources in your food additive company and since these resources tend to be in short supply and difficult to obtain externally, your company purchased and implemented an enterprise resource planning software package. This ERP system interfaces with the scheduling software and knowledge management system. It also changed over to critical chain scheduling. This means the company:

    a. Recognizes now that the next step is resource competency profiles
    b. Can tell when drum resources will be required
    c. Emphasizes resource productivity as a way to reduce overall costs
    d. Is using job shadowing as a means for others to acquire expertise of these SMEs

97. Assume you work for a pharmaceutical company. You are to recommend one new component to be optimized in the company's portfolio. You only can select one because of resource constraints. You can select Project A to develop a web site, Program A to enter the music video market, Program B for a major nationwide tour, or Project B for new T-shirts. You have net present value (NPV) data available to assist you in making your recommendation as follows:

Project A NPV	Program A NPV	Program B NPV	Project B NPV
5% = 3,524	5% = 2,201	5% = 6,400	5% = 3,055
10% = 2,901	10% = 2,254	10% = 3,275	10% = 2,857
15% = 1,563	15% = 1,632	15% = 1,679	15% = 1,125

You recommend:

a. Project A
b. Program A
c. Program B
d. Project B

98. Assume you are the Service Information Department's Director in your company that manufactures tractors and other farm equipment. You manage components such as services to dealers and development of operational and support manuals. You lack awareness at times as to when a new product is about to be available and then must provide needed resources to support it immediately. Time to market is critical in this industry. In terms of portfolio risk management, your primary risk concerns involve:

   a. Application of best practices
   b. Processes to support change
   c. Shifting portfolios
   d. Technological advances

99. Working to establish a portfolio management practice in your ice cream company, with franchises throughout your country, you are using portfolio process assets, organizational process assets, and best practices from bench-marking forums and in published reports. You also are evaluating various external environmental factors, and in this scenario, one to consider is:

   a. Work instructions
   b. Calendars
   c. Component proposals
   d. Workmanship standards

100. An internal auditor recently was asked to review all the processes under way in the organization including those at used in program, project, and portfolio management. In the audit team's report, one item noted was the impact of the project processes to those at the portfolio level. This item is significant as:

   a. All three must be aligned to the organization's vision and strategic goals
   b. There is a lack of transparency impacting the selection process
   c. Project managers lack training in the portfolio process
   d. Greater visibility into the prioritization process is needed

101. Each month, your Finance Department issues a series of reports on the sixth day of the month. This is especially helpful to you as you are the portfolio manager and know these data are of interest to the members of the Portfolio Review Board. When the company implemented portfolio management, it decided to hold Board meetings on the 10th day of each month. To assist in this process, you are using a communications calendar as it:

   a. Shows portfolio communication dependencies
   b. Describes stakeholder reporting requirements
   c. Emphasizes value-added impacts to the portfolio's performance
   d. Complements the portfolio strategic plan

102. Your retail foods company has numerous people who are in a union, and your management is concerned that because of the economic downturn in your country, that there could be a strike for higher wages that could last for some time. The Portfolio Oversight Group realizes it must continue to add new components to its mix to remain competitive especially as special-ized food stores are opening close to your stores' locations but is reluctant to do so given the union situation. As the portfolio manager as you work to optimize the portfolio, you are striving to decrease their concerns by using:

    a. SWOT analysis
    b. Scenario analysis
    c. Competitor analysis
    d. Probability analysis

103. The ROI on your overall portfolio has led your management in your manu-facturing company to question whether streamlining is needed to remain profitable before approving new components to be part of the portfolio. It is concerned as to whether its factories are operating efficiently and/or whether some are not needed at all. To assist in their decision making, you have been charged with conducting a:

    a. Sensitivity analysis
    b. Gap analysis
    c. Capacity analysis
    d. Dependency analysis

104. Assume you have been the portfolio manager, a new position in your Water Resources Department, for a year. A Portfolio Oversight Committee meets monthly, and you have set up categories for your components to best rank or score them and for resource allocation. Although groundwater is plenti-ful, the Climate Change Interest Group has been lobbying the Department to provide greater emphasis on sustainability, and this group is lobbying across your country. Your Oversight Committee requested that a sustainability cat-egory be added to the current list. You need to:

    a. Update the portfolio strategic plan
    b. Update the portfolio management plan
    c. Update the portfolio roadmap
    d. Review existing components in the portfolio to see if they should be in this category

105. Assume your organization is in the underground sprinkler system business. It is striving to be the market leader in this field in your State, and a major compliant from customers of sprinkler systems is the need to repair them frequently. As your company strives to be the market leader in the field, it is offering a new series of products as part of its new product pipeline. This pipeline provides:

    a. Stakeholders with confidence that the company is continually looking at new products to continue to thrive
    b. A leading indicator of future sales potential
    c. An approach to show the expected net benefits of the overall portfolio
    d. A balanced scorecard approach to measure performance

106. Assume you have identified five categories for the risks in your portfolio: organizational risk, performance risk, market risk, image risk, and financial risk. For each one in your portfolio risk management plan you also want to assess how you would measure it. In this measurement section of the plan, you also can include:

    a. When portfolio risk management will be performed
    b. The various risk management activities
    c. Team member involvement
    d. Risk appetite

107. In using a bubble diagram, there are a number of different variations to consider. In your company, you decide to set one up that shows business units, investments in various categories, investments in the business unit, and the components by category. You also are using the size of the bubble to show the importance of the component. Such an approach is effective as it:

    a. Accurately depicts the data that are used
    b. Ensures there is only one representative diagram
    c. Provides an overall view of the portfolio
    d. Shows when the components will be completed

108. You are the portfolio manager in your defense contracting firm. Your Portfolio Review Board has approved the strategic plan you prepared and now has asked you to prepare a portfolio management plan in order to describe:

    a. Your responsibilities as the Portfolio Manager
    b. A chronological view of the high-level strategic direction
    c. The alignment of the portfolio objectives with the strategic plan
    d. Management's intent to prioritize the work to meet the strategic objectives

109. Assume you work for the leading high-end department store in your country. It is considered to be the most fashionable store, and people enjoy the entire shopping experience. Many have personal shoppers to accompany them. However, earlier this year, many customers were victims of credit card fraud, and sales have decreased. You are sponsoring a new program so robots will be available at the time of checkout and will ensure no one is around when the customer leaves. The robots are to be fun for customers to see and equipped so that security breaches cannot occur, providing a high level of confidence for shoppers. The program has an aggressive time table and is to be complete in three months, making it the leading priority in the portfolio. As the program's sponsor, your expectations are to know about:

a. All developments of consequence
b. All changes in the portfolio
c. Portfolio milestones, risks, cost, and schedule
d. Portfolio changes, risks, and issues

110. Assume your organization has an organizational strategy, a defined mission and vision, and strategic goals and objectives, which are reviewed at least annually. It is implementing portfolio management, and it therefore needs key organizational enablers to assist in this process, which include:

a. A Portfolio Management Office
b. Metrics or Key Performance Indicators
c. Knowledge, skills, and competency profiles of staff members
d. Defined roles and responsibilities

111. Your transit authority, which serves a major metropolitan area in your country, has had a series of late completions of its programs and projects, numerous customer complaints regarding poor service, and complaints from its staff about being overworked. The CEO was fired, and a new CEO from a new product development organization, in which strict governance processes and stage gate reviews were held, was appointed. He then selected you as the Portfolio Manager and has set up a Portfolio Review Board. You are now collecting data on existing components. As you do so, you realize:

a. Benefits should be collected for each component
b. The data should be quantitative
c. Each program or project manager should be interviewed
d. It will be revised several times

112. As you began to determine the information needs of your stakeholders as part of preparing the portfolio communications management plan, you began with sponsors and governance board members. As you talked with them, you then learned about others who were influential in portfolio management and gathered additional information. This means:

    a. An increasingly large number of stakeholders want to be involved in the process
    b. Managing stakeholder expectations will consume the majority of your job
    c. You need to update the portfolio management plan
    d. You need to update the portfolio strategic plan

113. Having worked for eight different agencies over your government career, you know change is constant. Some programs you worked on at the time seemed as if they would last forever, but then some type of strategic change occurred, and they were terminated. In one government agency, where you were the portfolio manager, the government decided to make it one that was quasi-governmental, meaning some funding would be provided by the government, but the agency would need to find additional funding from other sources or reduce the components in the portfolio dramatically. This meant:

    a. Continual review and approval by stakeholders rather than only at stage gates
    b. Consistent communication to all stakeholder groups
    c. A separate plan to address compliance requirements with this new legislation
    d. A defined change control structure

114. Each organization is unique, and the same is true in how portfolio management is practiced. However, assume there is a strategic change to an organization that requires regulatory approval before its products can be commercialized. This new change now requires yet another audit before the product can be generally recognized as safe. Passing this new audit will add time to each schedule as it is unsure how quickly they can be done. This is a major strategic change, which affects much of the work under way, and shows that:

    a. It is necessary to work closely with the strategic planning staff as the portfolio manager
    b. The portfolio communications and risk plans also require updates
    c. Buffers need to be added to each program and project schedule because of these audits
    d. The prioritization criteria require change

115. A number of different portfolio risks affect executives, functional managers, the portfolio manager, and program and project managers. One example is:

    a. Information about portfolio components
    b. Organizational integrity
    c. Impact on the organization's strategy and objectives
    d. Quality of data about the portfolio

116. As the portfolio manager, you recognize the next meeting of the Portfolio Oversight Group will be contentious as regulatory mandates have led to the need for new components plus there are proposals for other components that may lead to new breakthroughs for this pharmaceutical company. However, you also know there are resource shortfalls especially among the scientific and engineering staff. To support the case to acquire additional resources, you decided to track over time the impact of optimization decisions on resource capacity. This means you:

    a. Used sensitivity analysis
    b. Performed resource leveling
    c. Prepared a detailed forecast to show resource supply and demand
    d. Assessed resource bottlenecks

117. As you work on a detailed portfolio performance plan to complement the other documents that are necessary to implement portfolio management efficiently in your organization, you found the portfolio management plan to be useful as it:

    a. Includes information on organizational culture
    b. Provides resource work calendar data
    c. States stakeholder expectations and requirements
    d. Includes financial-related policies

118. As the portfolio manager, you and your team provide monthly status reports on the overall health of the portfolio to internal stakeholders and shorter reports to key external stakeholders that lack confidential data. As you prepare these reports a best practice to follow is to:

    a. Set up a series of reports based on your stakeholder classification
    b. Filter each report since there are more than 1,000 components for an overall summary
    c. Have a single format for each component
    d. Consolidate the reports that are submitted

119. Assume you have determined the needed structure for your portfolio, and the proposed areas and categories of work have been agreed upon by the members of the Portfolio Review Board. You now need to review and update the portfolio strategic plan as a result of this work. Not to be overlooked in this process is updates that involve:

    a. The key and major stakeholders
    b. Dependencies
    c. High-level scope
    d. Benefits

120. Each time a new component is approved by the Portfolio Oversight Group, it means new stakeholders will have a greater interest in portfolio management in the organization. This means as you work to manage which stakeholders receive which information, you need to review the:

    a. Communication requirements analysis
    b. Portfolio management information system
    c. Dashboards
    d. Portfolio

121. When your government agency implemented portfolio management as mandated by the President in the previous administration, each agency was required to have a Portfolio Oversight Committee as a way to reduce expenditures on meaningless programs and projects. The Committee was to make decisions and meet quarterly. It was effective at first, but then became the 'fad' of the administration and was not a mandate when the new President was elected. This means:

    a. The portfolio management plan may need revision
    b. Each agency should use resources dedicated to portfolio management to other initiatives
    c. The Committee should remain but meet at the beginning of the budget cycle
    d. The emphasis should shift to 'doing programs and projects right'

122. Assume your organization desires to add more rigor to its selection process than just comparing one possible component to another and has added some additional criteria. Consider the table below:

Components	Criterion 1		Criterion 2 Probability of Success		Criterion 3		Criterion 4		
	ROI	Rank	Result	Rank	Degree of Importance	Rank	Payback Period	Rank	Priority
Project 1	12.0		2.1		1.5		4.6 years		
Program 1	14.0		18.9		4		2.5 years		
Project 2	15.5		8.45		2		3 years		
Operational Activity	19.0		5.95		1 (– –)		1.3 years		
Program 2	16.0		8.8		5 (++)		2 years		
Project 3	10.0		5.4		3		5.2 years		

Based on the data, you would recommend:

a. Project 1
b. Operational Activity
c. Project 2
d. Program 2

123. Assume you work in the beef industry. You have performed a risk assessment and have identified five high priority risks to the portfolio: regulatory changes, random inspections, disease, poor employee morale, and technology changes. Each of these risks requires additional analysis before extensive monitoring will begin to watch for it. As part of your analysis, especially since millions of pounds of beef were recalled that affected your company and five others, you decided to review the portfolio over the past three years for changes and:

a. Prioritize risk lists
b. Focus on sales forecasts
c. Determine trends
d. Assess the frequency of meeting cost objectives

124. You are considering proposing a new component as you work to optimize the portfolio. This component is to design the next generation refrigerator that also can serve as an oven and a dishwasher so there is only one large appliance in one's home. It will use state-of-the art technology and will be offered at a reasonable cost. It is designed to be attractive and will not require much space. To support your recommendation, you decide to conduct:

    a. An analysis of the expected benefits from the component
    b. A feasibility study
    c. An analysis of competing efforts under way in the company
    d. A SWOT analysis

125. Many categories are useful for portfolio management, and it is a best practice in determining them that they focus on the key areas that contribute to overall business value. Assume your company manufactures anti-bacterial soap products and has been doing so for over 40 years. It also focuses on developing new soap products to appeal to different consumers. While regulatory/compliance is a key category for your company's components, another is:

    a. Foundational
    b. Urgency
    c. Support strategic goals
    d. Resource requirements

126. The portfolio manager relies on an up-to-date list of components in the portfolio. With an understanding of the components in the portfolio, it leads to an effective communications approach as it:

    a. Makes sure all component managers receive all portfolio-related information
    b. Enables the development of a standardized portfolio communications plan
    c. Can standardize communications at the portfolio level
    d. Links portfolio performance with stakeholder information requirements

127. As you work on the structure for components to be in your portfolio charter, since this is a new approach in your pharmaceutical company, you are setting up various assumptions for the Portfolio Review Board using different combinations of potential components and new components. The objective is to determine different portfolio scenarios. One approach to use is:

    a. Cost-benefit analysis
    b. Prioritization analysis
    c. Interdependency analysis
    d. Scenario analysis

128. You are managing a new product designed to transform your company so it gains market share and provides a less expensive and easier to use approach than Cloud Computing. Your executives set up a collocated team for this project, which is located off site from the company's main campus. Everyone on the team has signed a non-disclosure agreement so competitors do not find out about this product. It is ranked number one in the company's portfolio. As the project manager, you expect to know about:

    a. Concerns of the Portfolio Review Board members
    b. Changes, risks, and issues
    c. Benefits and outcomes toward strategic goals
    d. Financial standing of the company

129. Your company has a terrific opportunity to bid on a contract from the U.S. Department of Defense. While your firm has the required expertise to do the work, and the campaign manager has an excellent working relationship with the customer, there is a requirement that to bid on this project, the company must be CMMI Level 3 certified. Your company is only certified at Level 2. This requirement is one that represents a negative risk even though the customer has been involving the campaign manager in discussions about the project for two years. Everyone was surprised by this clause in the Request for Proposal (RFP). The best course of action is to:

    a. Attend the bidders conference and ask that this clause be removed
    b. Note in the proposal that the company is Level 2 certified and expects to be Level 3 certified by the time of the contract award
    c. Form a consortium with another firm that is Level 3 certified
    d. Work with a company that is Level 3 certified as a subcontractor

130. Each Portfolio Review Board meeting tends to address certain issues. While some can be resolved during the meeting, such as whether or not to reprioritize a portfolio component because a key resource is unavailable, others require more time such as the need to acquire technology. As the portfolio manager, you believe it is useful to present at these meetings:

    a. An issue log
    b. An issue aging chart
    c. A list of both outstanding risks and issues to resolve
    d. Those issues that involve more than one portfolio component

131. Your health insurance company has set up its portfolio into five different categories: research and development, IT, Medicare, government health insurance, and non-government health insurance. Funding is allocated yearly to each of these six categories. As the portfolio manager at the enterprise level, you:

    a. Ensure such allocations are reflected in the portfolio's strategic plan
    b. Meet with the CFO and determine these allocations when the budget for the fiscal year is being prepared
    c. Meet with the managers of the five portfolios once the budget allocations are known
    d. Use your existing inventory of components in the portfolio and in the pipeline to determine funding allocations

132. The two major purposes of the portfolio charter are to authorize the portfolio manager to:

    a. Execute the strategic plan and document roles and responsibilities
    b. Staff a Portfolio Management Office and to set up a governance structure
    c. Establish portfolio management processes and procedures
    d. Execute portfolio processes and apply resources to components

133. Recently the portfolio manager in your pet food company left. You applied for the position and were selected. Fortunately, the previous portfolio manager maintained a decision register from Portfolio Review Board meetings. This decision register:

    a. Should enable sufficient information as to why a component was selected, deferred, or rejected
    b. Provides data as to which members of the Board supported each decision and those who were not in favor of it
    c. Should be part of the organization's knowledge repository
    d. Is a portfolio process asset

134. Your investment company has practiced portfolio management building on concepts from Markowitz since it began a corporate focus on program and project management 30 years ago. Since then, a number of portfolio process assets have been collected and retained. As the portfolio manager, you recognize their importance especially as you work to manage the available portfolio information, but you need to assess their usefulness based on:

    a. Results from your communications requirements analysis
    b. Insight from your stakeholder analysis
    c. Available communications methods
    d. The communications matrix

135. Similar to a risk management plan for a program or project at the portfolio level, this plan shows how portfolio risk management activities will be set up and managed. Not to be overlooked as a part of this plan is:

    a. References to corporate risk guidelines
    b. The specific risks that have been identified
    c. Portfolio reports
    d. Portfolio issues

136. You have prepared a resource schedule to use before Portfolio Review Board meetings as a form of a resource capacity and capability analysis. Since you work in a highly scientific organization, intellectual capital is a key area of concern. This approach is useful because it:

    a. Supports resource leveling
    b. Shows impacts of resource optimization decisions
    c. Indicates resource bottlenecks
    d. Combines and details forecasts of ongoing supply and demand

137. One advantage of the roadmap is that it communicates to stakeholders:

    a. The scope of the portfolio
    b. The link between organizational strategies and portfolio management
    c. The mapping of benefits to components
    d. The resources required to support the components and when they may be needed

138. As the portfolio manager for the Federal Railroad Administration, recently there was a strategic change to the portfolio because of the need to develop and implement new safety regulations to avoid train derailments because of the need to travel at high speeds when transporting oil and hazardous materials. To develop these regulations, which are considered a high priority by the Administrator and outside special interest groups, the portfolio requires new components to also ensure the regulations are implemented and followed. This means as the portfolio manager you should:

    a. Engage stakeholders through targeted messages
    b. Design reports that differ from those for other components since these components have high visibility
    c. Announce this change through a press release
    d. Prepare a report based on key risks and issues with these components

139. Your pharmaceutical company implemented a sophisticated portfolio management process seven years ago and purchased an expensive software tool to help in prioritizing projects. However the process turned out to be too cumbersome, and your CEO hired a leading consulting firm to make recommendations to simplify it. Now each division prepares a simple spreadsheet of its top six portfolio components as shown below:

	Project A	Program A	Operational Activity A	Project B	Project C	Program B	Rank
Project A		1	1	1	1	1	5
Program A	0		0	1	1	0	2
Operational Activity A	0	1		1	1	0	3
Project B	0	0	0		1	0	1
Project C	0	0	0	0		0	0
Program B	0	1	1	1	1		4

This approach is useful as it:

a. Uses different criteria based on component type
b. Shows the highest score
c. Presents a measure of success based on the ratings
d. Weights criteria by importance

140. Assume you are preparing a portfolio risk management plan. You are talking with key stakeholders as well as reviewing existing artifacts. One that is easy to overlook is:

a. Risk processes
b. Scoring techniques
c. Risk interdependencies
d. Vision statement

141. From your research in organizations that have been successful in portfolio management, as you are doing some benchmarking as you implement it in your organization, you find many organizations decide to do so as there are more programs and projects under way than available resources to execute them. Assume you are now working to develop the portfolio charter. In doing so, it is useful to review existing portfolio process assets. One that should not be overlooked is:

   a. Skill set limitations
   b. Organizational variables
   c. Stakeholder relationships
   d. Interdependencies

142. Assume you are working with the Director of the EPMO to develop a complete inventory of the work under way in the company. Such an inventory has not been developed previously. You characterize your work as one that is:

   a. Immature but evolving
   b. Incomplete as some people do not wish to disclose their work as they fear it could be terminated
   c. As complete as possible given the EPMO's support
   d. Necessary to develop the portfolio management plan

143. While your social media company is new to portfolio management, the executive team is committed to its implementation, and you are pleased to be the portfolio manager since you have your PfMP. You have prepared your portfolio strategic plan, roadmap, and charter. The next logical step is to:

   a. Set up the governance structure
   b. Convene the first meeting of the Portfolio Review Board
   c. Prepare the portfolio communications plan
   d. Prepare the portfolio management plan

144. Working for the leading producer of containers for ice cream, your company is facing more competition. When other companies were producing square containers, your company was the only one producing round ice cream containers. The competitors are now doing the same. You have been hired as the portfolio manager to set up portfolio plans and a process to follow to add new products and services to the company's product line. You are preparing your portfolio management plan and want to involve as many people in your company as possible so they feel they contributed to the plan and are familiar with its objectives. You decided to:

    a. Set up cross-functional focus groups
    b. Interview a representative from each department
    c. Use brainstorming
    d. Use surveys

145. Managing stakeholder expectations is critical to success regardless if one is the portfolio manager or is working on programs, projects, or in functional areas. As a portfolio manager, establishing working relationships with stakeholders is fostered through:

    a. Involving them in preparing policies and procedures
    b. Identifying and classifying them
    c. Preparing and following a stakeholder expectations plan
    d. Preparing and following a transparent communications strategy

146. In preparing a list of all the work under way in your new product development organization, a number of the active components have key interfaces and dependencies with other components. These dependencies must be tracked, and one way to do so is to:

    a. Show them in the portfolio management plan
    b. Use Key Performance Indicators
    c. Use the roadmap
    d. Document the dependencies in each component's program or project plan

147. Assume since you are the portfolio manager for your food and beverage company, you are preparing a portfolio risk management plan. This plan is needed as there are numerous risks now from external consumer groups about the nutritional value of some of your products plus the company must comply with an increasingly large number of regulations by the Agriculture Department and the Food and Drug Administration. There are also numerous inspections of the quality of the products. Your Portfolio Review Board asked you to prepare this plan in order to:

    a. List the negative and positive risks
    b. Provide an approach to assess risks in its existing components
    c. Provide an approach to assess risks in proposed components
    d. Determine the number of high risk, high return components to pursue

148. Assume your organization is committed to a management-by-projects culture. Its leaders in approving components to be part of the portfolio also consider the structure required to manage each one, and based on the structure, resource allocation then is a determining function. Its top five components in the portfolio are ones that operate in a projectized structure meaning that:

    a. External resources can be acquired easily since a contracting officer is part of the team
    b. The capabilities and competencies of the people working on these five components are known
    c. The emphasis is on ensuring resources supporting these components are competitive and cannot be easily purchased
    d. Resources are unlimited

149. As the portfolio manager, you need to assign performance reviews and reporting roles and responsibilities. Assume you are working for the leading professional association in project management, and it has six distinct business units. Each of these business units has programs, projects, and operational activities under way. You are assigning these roles and responsibilities as you:

    a. Implement your portfolio management plan
    b. Set up a process for an oversight committee at the association at the highest level
    c. Implement organizational structure analysis
    d. Integrate the various portfolio plans for ease of reporting on progress

150. Assume you are charged with implementing portfolio management in your City's Transit Authority at the enterprise level. This is a coveted assignment, and you are pleased you were selected as its lead as you saw its effectiveness while doing some benchmarking with other transit authorities in your country. You have prepared a communications plan that the CEO personally sent to stakeholders in the Authority. Your next step is to:

    a. Begin drafting the processes to be followed
    b. Convene a meeting of the Portfolio Review Board
    c. Set up training for staff and stakeholders
    d. Document external factors that will affect the implementation

151. After a recent government shutdown, your executives realized that some functions in your Agency were not missed, but others needed to be added to better support the strategic objectives. As the portfolio manager, you need to prepare an updated portfolio strategic plan. In order to do so, you should:

    a. Prepare an analysis to support reallocation of resources
    b. Determine if the Agency has the needed competencies for the new functions or whether contractors will be needed
    c. Show where there are gaps in focus, investment, or alignment
    d. Revisit the prioritization process to guide you as you update this plan

152. Recently, the Portfolio Review Board met and optimized the portfolio because of some strategic changes. As it was balanced, some components were added, and others were terminated. This optimization meant updates to the portfolio management plan also were needed. An example of an item to update is:

    a. Ways to maintain a balanced portfolio
    b. Organizational areas in the portfolio
    c. Portfolio guidelines
    d. Ranking criteria

153. As you have worked to improve the quality of information on the portfolio to internal and external stakeholders, you found you needed to update reports and plans. You then reviewed the portfolio process assets and need to update:

    a. Dashboards
    b. Trend analyses
    c. Component status reports
    d. Meeting minutes

154. Portfolio risks may be identified by anyone in the organization, and those organizations with mature portfolio management processes and involvement from stakeholders encourage people to openly communicate about risks. This open approach can promote positive risks and minimize the impact of negative risks. The Procurement and Contracts Director recently expressed concern about an excessive reliance on a specialized training provider. This is an example of a:

    a. Positive or negative risk
    b. Negative risk
    c. Structural risk
    d. Reputation risk

155. Assume your company, established almost 80 years ago, is well-known for its ice cream products and has franchises throughout the county. But, the government is focusing on ways to try to eliminate obesity in both adults and children. Sales are decreasing, and some of the franchises are closing. The company decided to change its strategy to continue with the ice cream products but to enter the nutritional cereal market as well. As the portfolio manager you realize this is a major strategic change and evaluated the existing portfolio and its components against the updated strategy as:

    a. Many existing components will no longer be needed
    b. Descriptors and categories may need change
    c. A weighted ranking and scoring technique should be used
    d. Both quantitative and qualitative analyses are required

156. Since your government agency is undergoing a strategic change and will now use contractor staff for all operational activities beginning with the new fiscal year in two months, as the portfolio manager, you must update the portfolio strategic plan. Among the various items in the plan, you need to update which of the following sections:

    a. Communications management
    b. Stakeholder engagement
    c. Benefits
    d. Performance metrics

157. Assume your 500-person consulting company has taken time to develop and maintain competency profiles for each staff members. It also has set up resource pools based on job title and uses earned value management. Your company recently won two major contracts, and you were the proposal manager for both of them. As you assess resource requirements:

    a. You need to use external consultants for some of the work as staff are allocated on other projects
    b. Reports on resource availability should be useful
    c. Your first step is to review the staff competency profiles
    d. You need to see if any projects recently have been completed

158. Implementing portfolio management is a culture change and takes time to do so effectively. However, it is necessary to describe how the portfolio will deliver value to the organization, which is stated in the:

    a. Portfolio performance plan
    b. Key performance indicators
    c. Portfolio benefits realization plan
    d. Portfolio charter

159. Assume you are putting together for the Portfolio Review Board several options for consideration of potential components and current components. You are using an approach with different probabilities to determine outcomes and EMV. Which of the following would you recommend realizing Components A and B are new, while C and D are in progress?

		Component A		Component B		Component C		Component D	
	Probability	Outcome	EMV	Outcome	EMV	Outcome	EMV	Outcome	EMV
1	60%	$15,000	$9,000	$13,000	$7,800	$20,000	$12,000	$12,000	$7,200
2	25%	$17,000	$4,250	$15,000	$3,750	$12,000	$3,000	$10,000	$2,500
3	15%	$20,000	$3,000	$15,000	$2,250	$10,000	$1,500	$8,000	$1,200

    a. Component A
    b. Component B
    c. Component C
    d. Component D

160. Having managed a major program in your rice producing company, you found it valuable to maintain a program risk register. Now assume you are the portfolio manager. Yesterday there was an unexpected risk to the portfolio's key component as the product had to be recalled because of safety issues. An investigation into the root cause is under way, and before the product can return to the market, an inspection by a regulatory official will be necessary. You have been using a risk register to track all identified risks until they are closed and have added this risk to it. One of the benefits of the risk register is that it:

    a. Ensures all risks are documented

    b. Assigns an owner to each risk

    c. Defines the frequency needed for risk identification

    d. Contains probability and impact information for the risks in the register

161. Assume your government agency set up a Portfolio Oversight Committee chaired by the head of the Agency. The other members are the Directors' of the Agency's departments. You are its Knowledge Management Officer and are responsible for attending all the meetings that are held in order to:

    a. Provide an objective perspective as to which new components to pursue

    b. Ensure compliance with standards

    c. Communicate decisions through the knowledge repository

    d. Provide information if something similar was considered in the past but was deferred

162. When you began your job as the portfolio manager for your dairy cooperative, your executives suggested it be pilot tested in one business unit and selected the animal drug residue unit. It was a success, and in two years, you received a promotion as portfolio management then was implemented across the cooperative. While in the animal drug residue unit, you collected data on spreadsheets on components that you analyzed when meetings of the Oversight Group in the business unit were held. Now your plan is to:

    a. Improve exception reporting cycles

    b. Acquire an automated PMIS

    c. Ensure stakeholders receive targeted information based on the communications requirements analysis

    d. Review dependencies in the portfolio along with stakeholders' time-sensitive needs

163. Working for a government agency that performs some vital functions but one in which none of these functions is considered critical to national security or the health and well being of its citizens, as the portfolio manager you are accustomed to having to 'do more with less'. Another round of budget cuts is expected. This means that:

    a. Resource limitations are expected as many will be offered options for early retirement
    b. Strategic opportunities will require approval from those in the executive branch of the government outside of the agency
    c. Overhead functions such as the EPMO and staff member continual training will be eliminated
    d. Allocation of funds to different types of initiatives and their contribution to strategy may require revision

164. Assume you are the portfolio manager for a global *Fortune* 50 company specializing in the production of golf balls since golf is so popular especially with the aging population. The company has other products under way or planned so it remains the market leader in this area. The Portfolio Review Board meets monthly, and at each meeting, it reviews the portfolio in order to:

    a. Address impending risks and outstanding issues
    b. Consider funding allocations
    c. Evaluate if the portfolio's benefits align with objectives
    d. Ensure agreement on the overall timeline for components

165. Since robots and small drones are being used more frequently in very dissimilar ways, your company, rather than implementing new proposals in this area since it is the market leader in dirigibles, immediately stated component proposals for new investments are first to be approved in governance board meetings. This means as you manage portfolio information that:

    a. Communications are built around this policy
    b. A focus on strategic program and project data is required
    c. Governance board members must be selected carefully to ensure a transparent process
    d. Risks and issues require significant attention

166. When the portfolio strategic plan is prepared, and each time it is updated, it is prepared by using:

    a. Organizational and portfolio strategic assets
    b. Organizational strategy and objectives
    c. Organizational vision and mission
    d. Portfolio benefits and business value expected

167. Criteria to ensure alignment to strategic goals is necessary in order to:

    a. Ensure that only approved programs and projects are in the portfolio
    b. Determine an overall score for each component
    c. Ensure a strong link between programs and projects and the strategic plan
    d. Indicate if there are gaps within the portfolio

168. Working to manage the value of the portfolio, you worked with stakeholders in the seven business units of your energy company to estimate various outcomes considering the success criteria of the portfolio. The Portfolio Review Board has five members. One is the CEO of the company, and another member is the CFO. The other three members are comprised of the Directors of the business units and rotate every six months. You found five of these directors are risk adverse, and since you report to the CEO, you realize his goals are to venture into new markets with new technology to increase short-term profits. Since the CFO also is risk adverse, the Board basically is risk adverse. You met with the CEO, and he proposed a new approach to select Board members, and you are to implement it immediately. This means you need to:

    a. Update the portfolio management plan
    b. Meet with those Board members who will no longer be represented
    c. Ask the CEO to meet with affected stakeholders
    d. Update the portfolio strategic plan

169. Your organization has a Center of Excellence in project management, which has evolved over the past eight years to also cover programs and portfolios. Its success has been noted by others, and it is often selected for benchmarking. It received a PMO of the Year Award for its achievements. In terms of portfolio management, it:

    a. Serves as the oversight committee to approve, defer, or terminate components
    b. Has the responsibility to develop, implement and maintain the prioritization model
    c. Defines the portfolio processes and procedures
    d. Provides a way to share or optimize scarce resources

170. Being a new portfolio manager, assume you have been asked to conduct a strategic alignment analysis because two of the organization's strategic goals now are obsolete because of a merger with a smaller competitor. One step to follow in conducting such an analysis is to:

   a. Use the 'as is' and 'to be' state and graphically display the components by application areas
   b. Compare each component in the top ten in the portfolio to one another through a weighted ranking approach
   c. Start with the actual value of each component and compare it to the proposed expected value
   d. Determine the current allocation of funds to each application area and evaluate the need for changes based on the merger

Answer Sheet for Practice Test 2

1.	a	b	c	d
2.	a	b	c	d
3.	a	b	c	d
4.	a	b	c	d
5.	a	b	c	d
6.	a	b	c	d
7.	a	b	c	d
8.	a	b	c	d
9.	a	b	c	d
10.	a	b	c	d
11.	a	b	c	d
12.	a	b	c	d
13.	a	b	c	d
14.	a	b	c	d
15.	a	b	c	d
16.	a	b	c	d
17.	a	b	c	d
18.	a	b	c	d
19.	a	b	c	d

20.	a	b	c	d
21.	a	b	c	d
22.	a	b	c	d
23.	a	b	c	d
24.	a	b	c	d
25.	a	b	c	d
26.	a	b	c	d
27.	a	b	c	d
28.	a	b	c	d
29.	a	b	c	d
30.	a	b	c	d
31.	a	b	c	d
32.	a	b	c	d
33.	a	b	c	d
34.	a	b	c	d
35.	a	b	c	d
36.	a	b	c	d
37.	a	b	c	d
38.	a	b	c	d

39.	a	b	c	d		61.	a	b	c	d
40.	a	b	c	d		62.	a	b	c	d
41.	a	b	c	d		63.	a	b	c	d
42.	a	b	c	d		64.	a	b	c	d
43.	a	b	c	d		65.	a	b	c	d
44.	a	b	c	d		66.	a	b	c	d
45.	a	b	c	d		67.	a	b	c	d
46.	a	b	c	d		68.	a	b	c	d
47.	a	b	c	d		69.	a	b	c	d
48.	a	b	c	d		70.	a	b	c	d
49.	a	b	c	d		71.	a	b	c	d
50.	a	b	c	d		72.	a	b	c	d
51.	a	b	c	d		73.	a	b	c	d
52.	a	b	c	d		74.	a	b	c	d
53.	a	b	c	d		75.	a	b	c	d
54.	a	b	c	d		76.	a	b	c	d
55.	a	b	c	d		77.	a	b	c	d
56.	a	b	c	d		78.	a	b	c	d
57.	a	b	c	d		79.	a	b	c	d
58.	a	b	c	d		80.	a	b	c	d
59.	a	b	c	d		81.	a	b	c	d
60.	a	b	c	d		82.	a	b	c	d

83.	a	b	c	d
84.	a	b	c	d
85.	a	b	c	d
86.	a	b	c	d
87.	a	b	c	d
88.	a	b	c	d
89.	a	b	c	d
90.	a	b	c	d
91.	a	b	c	d
92.	a	b	c	d
93.	a	b	c	d
94.	a	b	c	d
95.	a	b	c	d
96.	a	b	c	d
97.	a	b	c	d
98.	a	b	c	d
99.	a	b	c	d
100.	a	b	c	d
101.	a	b	c	d
102.	a	b	c	d
103.	a	b	c	d
104.	a	b	c	d

105.	a	b	c	d
106.	a	b	c	d
107.	a	b	c	d
108.	a	b	c	d
109.	a	b	c	d
110.	a	b	c	d
111.	a	b	c	d
112.	a	b	c	d
113.	a	b	c	d
114.	a	b	c	d
115.	a	b	c	d
116.	a	b	c	d
117.	a	b	c	d
118.	a	b	c	d
119.	a	b	c	d
120.	a	b	c	d
121.	a	b	c	d
122.	a	b	c	d
123.	a	b	c	d
124.	a	b	c	d
125.	a	b	c	d
126.	a	b	c	d

127.	a	b	c	d
128.	a	b	c	d
129.	a	b	c	d
130.	a	b	c	d
131.	a	b	c	d
132.	a	b	c	d
133.	a	b	c	d
134.	a	b	c	d
135.	a	b	c	d
136.	a	b	c	d
137.	a	b	c	d
138.	a	b	c	d
139.	a	b	c	d
140.	a	b	c	d
141.	a	b	c	d
142.	a	b	c	d
143.	a	b	c	d
144.	a	b	c	d
145.	a	b	c	d
146.	a	b	c	d
147.	a	b	c	d
148.	a	b	c	d

149.	a	b	c	d
150.	a	b	c	d
151.	a	b	c	d
152.	a	b	c	d
153.	a	b	c	d
154.	a	b	c	d
155.	a	b	c	d
156.	a	b	c	d
157.	a	b	c	d
158.	a	b	c	d
159.	a	b	c	d
160.	a	b	c	d
161.	a	b	c	d
162.	a	b	c	d
163.	a	b	c	d
164.	a	b	c	d
165.	a	b	c	d
166.	a	b	c	d
167.	a	b	c	d
168.	a	b	c	d
169.	a	b	c	d
170.	a	b	c	d

Answer Key for Practice Test 2

1. a. Organizational structure and organizational areas

    Through alternatives analysis an integrated view of the portfolio strategy can be prepared depicting multiple portfolio components in the areas of the organization; the organizational structure and areas therefore are described in more detail in the strategic plan.

    Portfolio Management Standard, pp. 44, 46

    Task 1 in the ECO in Strategic Alignment

2. c. Portfolio, sub-portfolios, programs, and projects based on organizational areas

    The structure aligns to the portfolio strategic plan. It also includes hierarchies, timelines, and goals for programs, projects, and operational activities.

    Portfolio Management Standard, p. 47

    Task 1 in the ECO in Governance

3. b. Using trade-off analysis

    Investment choices are a tool and technique as part of quantitative and qualitative analysis used in the Develop Risk Management Plan process. Trade-off analysis is one approach to consider since it determines the effect of changing at least one factor in the portfolio.

    Portfolio Management Standard, p. 127

    Task 2 in the ECO in Risk Management

4. d. Key stakeholder expectations and communication requirements

    For strategic change to be effective, key stakeholders need to be part of the process to ensure the change is accepted. The portfolio manager, while managing the strategic changes, should consult the portfolio strategic plan and consider how best to handle stakeholder expectations and communication requirements.

    Portfolio Management Standard, p. 53

    Task 5 in the ECO in Performance

5. a. Relationships among them

Stakeholder analysis is used to prepare the portfolio communication management plan. The interests of each stakeholder, their level of influence, and the relationships between them require consideration during the planning process.

Portfolio Management Standard, p. 110

Task 2 in the ECO in Communications

6. b. Compare strategic objectives

The first step in preparing a prioritization analysis is to compare the proposed components in terms of their support to the organization's strategic objectives.

Portfolio Management Standard, p. 51

Task 7 in the ECO in Strategic Alignment

7. a. Earned value

The progress measurement techniques used in managing portfolio value are the same as those at the component level; the most useful is earned value.

Portfolio Management Standard, p. 102

Task 9 in the ECO in Performance

8. a. The company's workload had increased

If components are not aligned and the organization lacks a successful evaluation and definition process, unnecessary and poorly planned components are in the portfolio, and as a result, the organization's workload increases.

Portfolio Management Standard, p. 64

Task 2 in the ECO in Governance

9. b. A project sponsor

Functional groups, such as production, manufacturing, finance, marketing, legal, information systems, human resources, administrative services, etc., may be stakeholders in the portfolio and also can serve as sponsors of portfolio components.

Portfolio Management Standard, p. 5

Task 6 in the ECO in Performance

10. c. Heat map

A heat map is a form of qualitative analysis used in the Monitor and Control Risks process. It serves as a point in time comparison of portfolio risks and looks at risks from the point of view only of exposure value.

Portfolio Management Standard, p. 133

Task 5 in the ECO in Risk Management

11. c. Components

Even though the acquisition was one in which similar products and services were offered, a change in the component mix will occur as some components will change if those of the acquiring company have different features that may be positive and should be added to the portfolio.

Portfolio Management Standard, p. 55

Task 7 in the ECO in Strategic Alignment

12. a. The high-level timelines for portfolio delivery

The charter formally authorizes the portfolio. By including high-level timelines for portfolio delivery in it, with stakeholder signoffs, it provides common agreement to meet stakeholder expectations.

Portfolio Management Standard, p. 39

Task 1 in the ECO in Performance

13. c. Resource histograms

Resource histograms are useful to graphically display over or under allocation of resources across the portfolio.

Portfolio Management Standard, p. 118

Task 5 in the ECO in Communications

14. b. Is viewed as a positive risk

This situation shows the need to improve product quality, and the company has invested in quality management, a positive risk, to be more effective than having to perform later corrective actions because of poor quality, a negative risk.

Portfolio Management Standard, p. 120

Task 1 in the ECO in Risk Management

15. d. Portfolio roadmap

The roadmap is useful as it summarizes strategic objectives, describes how the strategy may evolve, and summarizes milestones, dependencies, and risks.

Portfolio Management Standard, p. 66

Task 8 in the ECO in Strategic Alignment

16. b. Minimize portfolio risks

Portfolio reports are an input to the Provide Portfolio Oversight process. They are useful to evaluate components and decide actions to take to minimize risks and maximize portfolio benefits.

Portfolio Management Standard, p. 83

Task 1 in the ECO in Governance

17. a. Ensure all components are comparable

Key descriptors are established to provide a way to ensure all components are comparable and provide an approach to filter or eliminate any components since there are associated acceptance levels.

Portfolio Management Standard, p. 67

Task 2 in the ECO in Governance

18. b. Involve stakeholders

All relevant stakeholders require involvement and buy in in developing and executing the portfolio's components to best increase overall portfolio success.

PMI (2013) Managing Change in Organizations, p. 45

Task 7 in the ECO in Strategic Alignment

19. c. Provide information on progress and results

The portfolio manager recognizes how the portfolio relates to the organization's strategy and plays a key role in implementing the strategy by monitoring the initiation of initiatives in the plan and communicating progress and results.

Portfolio Management Standard, p. 15

Task 3 in the ECO in Communications Management

20. a. Recognizing portfolio dynamics

    In risk evaluation, the portfolio manager considers portfolio analysis including fiscal constraints, cost-benefit analysis, windows of opportunity, portfolio component constraints, and stakeholder dynamics.

    Portfolio Management Standard, p. 16

    Task 1 in the ECO in Risk Management

21. b. Have the portfolio roadmap available

    The roadmap, used as an input to optimize portfolio, shows the strategy for the portfolio's "to-be" vision to guide the optimization process.

    Portfolio Management Standard, p. 73

    Task 1 in the ECO in Performance

22. a. Reflect on the investments made or planned by the organization

    There are a number of activities involved in portfolio management, but the emphasis in performing them is to reflect first on the investments that the organization has made or plans to make.

    Portfolio Management Standard, p. 3

    Task 3 in the ECO in Strategic Alignment

23. a. Component A

    Component A will drop features if necessary in a trade-off situation. In the Optimize Portfolio process, market analysis is a tool and technique. The situation shows that this is a schedule-driven company and is most interested in time to market. This component is most aligned with the company strategy.

    Milosevic et al., pp. 75–76

    Portfolio Management Standard, p. 75

    Task 6 in the ECO in Performance

24. c. The portfolio management office is ideal to develop and collect the new metrics

    Change is constant, and therefore, the portfolio management office is ideal to develop new metrics as required and discontinue collecting others if they are not adding value.

    Portfolio Management Standard, p. 89

    Task 2 in the ECO in Performance

25. b. Strategic objectives must be optimized

   One item in the portfolio strategic plan is to ensure that strategic objectives can be optimized with available resources and risks. The portfolio should be validated against organizational strategy for consistency with evolving organizational mission, goals, or objectives and is the purpose of the portfolio strategic plan.

   Portfolio Management Standard, p. 39

   Task 1 in the ECO in Strategic Alignment

26. a. Stakeholder risk tolerances

   In addition in this plan concerning risks, it may contain roles and responsibilities, budgets, a schedule, risk categories, and definitions of probability and impact.

   Portfolio Management Standard, p. 124

   Task 2 in the ECO in Risk Management

27. a. Provide information to those in portfolio component support functions

   Such information is provided as well to stakeholders as the portfolio is updated, but those in support functions such as human resources, finance, the PMO, and procurement should not be overlooked.

   Portfolio Management Standard, p. 80

   Task 5 in the ECO in Governance

28. a. Information in a report used to authorize the portfolio

   Before the portfolio is authorized, as an input, various types of portfolio reports are reviewed and analyzed; this is an example of the type of data that could be part of a financial report.

   Portfolio Management Standard, p. 79

   Task 1 in the ECO in Performance

29. b. Describe the goals and objectives

   By starting with the plan's goals and objectives, the recipients will have a common understanding as to why the plan is important and how it can best meet their requirements.

   Portfolio Management Standard, p. 113

   Task 4 in the ECO in Communications

30. a. An interdependency analysis

The interdependency analysis is a tool and technique to aid in development of the roadmap. It is a variation of the interrelationship diagraph to show the relationship of the various dependencies among portfolio components.

Portfolio Management Standard, p. 51

Task 8 in the ECO in Strategic Alignment

31. c. Market-payoff variability

This investment choice approach focuses on pricing and sales forecasts. It depends on a number of marketing factors as it recognizes the impact of changing one or more factors, which may affect the portfolio itself or the portfolio's strategy.

Portfolio Management Standard, p. 127

Task 2 in the ECO in Risk Management

32. c. Lead to the potential change of the Department's strategic direction

Portfolio performance is monitored against organizational strategy and objectives with performance feedback used to provide input to the potential change of the organization's strategic direction.

Portfolio Management Standard, p. 9

Task 9 in the ECO in Performance

33. a. Organizational communication strategy

The strategic plan specifies the organization's tolerance for risks, communication strategy at the organizational level, and the organization's performance strategy.

Portfolio Management Standard, p. 60

Task 1 in the ECO in Strategic Alignment

34. b. Provide a copy of the portfolio management plan

This plan also describes the portfolio management approach and how it is defined, organized optimized, and controlled; it is an input to the Provide Portfolio Oversight process.

Portfolio Management Standard, p. 83

Task 4 in the ECO in Governance

35. c. The original portfolio charter

    The portfolio charter should be reviewed to ensure the charter and the portfolio remain aligned or whether an update to the charter is required.

    Portfolio Management Standard, p. 53

    Task 5 in the ECO in Performance

36. d. Manage necessary changes

    Since portfolio management embraces a new way of working with its emphasis on strategic initiatives, managing changes is an integral part of its planning process.

    PMI (2013) Managing Change in Organizations, p. 46

    Task 7 in the ECO in Strategic Alignment

37. c. Determining constraints from capabilities

    In this situation, training has been conducted, but its focus has been on tools and techniques. The analysis reveals training is lacking on interpersonal skills, critical to project management. It shows human resource capability and capacity will be a limiting factor.

    Portfolio Management Standard, p. 74

    Task 7 in the ECO in Performance

38. b. Use the portfolio and impact ratings as stated in the portfolio risk management plan

    Each organization has different methods and ratings for probability and impact. A risk owner should consult the portfolio risk management plan and follow the ratings in it.

    Portfolio Management Standard, pp. 128, 130

    Task 4 in the ECO in Risk Management

39. c. Use the human resources person on the Portfolio Review Board and ask him to navigate the negative impact on the people involved

    The human resources unit can identify skills, qualifications, and other competencies needed for success and can help acquire them. In addition it can help facilitate resource realignment and minimize any negative impacts resulting from organizational changes affecting the portfolio.

    Portfolio Management Standard, p. 13

    Task 5 in the ECO in Governance

40. b.   A funnel chart

Prepared in the shape of a funnel, these reports can show by component the amount of time one is working on the component and the labor rate. It then can be used to show resource efficiency. As an example, one can see if a resource is being used on too many components or if a person seems to be spending more than the allocated time on the component.

Portfolio Management Standard, p. 96

Task 7 in the ECO in Performance

41. c.   Components may support more than one of these areas

It is a best practice to set up organizational value areas to see the areas the components are most likely to impact. Components may support more than one area, and this approach can help determine resource allocation to best benefit the organization. It is important that each component support a value area in decision making.

Portfolio Management Standard, p. 85

Task 1 in the ECO in Strategic Alignment

42. c.   Use a dashboard with summary information in these areas

A variety of different types of dashboard reports can be prepared. They are useful to easily see multiple messages about the status of the portfolio in a graphical report.

Portfolio Management Standard, p. 117

Task 5 in the ECO in Communications

43. d.   You will use your contingency reserve and work aggressively to complete the component

Acceptance is a risk response strategy and can be passive or active. An active approach involves use of a contingency reserve in terms of time, money, and resources to handle the risk.

Portfolio Management Standard, p. 134

Task 6 in the ECO in Risk Management

44. d. Desired risk profile

   While the risk profile is detailed in the risk management plan, it also is considered in the portfolio management plan in the section on balancing the portfolio and managing dependencies. In balancing, a key area of emphasis is the ability to maximize the portfolio return according to the organization's predefined risk profile.

   Portfolio Management Standard, p. 63

   Task 4 in the ECO in Governance

45. c. Day-to-day processes

   Operations management and the involvement of its processes are essential for portfolio management along with program and project management as delivery of value is then realized through these day-to-day processes.

   Portfolio Management Standard, p. 7

   Task 8 in the ECO in Strategic Alignment

46. c. Program B

   Using NPV as a way to recommend components for optimization, a dollar one year from today is worth less than a dollar today. The more the future is discounted, or the higher discount rate, the less the NPV of the component. If the NPV is high, the component then is ranked high, leading to Program B.

   Milosevic, pp. 42–44

   Portfolio Management Standard, p. 74

   Task 7 in the ECO in Performance

47. c. Performance management plan

   As a subsidiary plan to the portfolio management plan, one purpose of this plan is to focus on resource management planning to best ensure the portfolio component mix is one that can generate maximum value.

   Portfolio Management Standard, p. 87

   Task 4 in the ECO in Governance

48. a. Update changes to the portfolio component lists

   This action is necessary based on the governance recommendations, and this update is an example of a portfolio process asset to update as an output of the Manage Portfolio Risks process.

   Portfolio Management Standard, p. 135

   Task 6 in the ECO in Risk Management

49. c.   Forecasting resource supply and demand

By forecasting resource supply and demand at the portfolio level, the Portfolio Management Office then supports the Enterprise Program Management Office so that resource demand can be further broken down into supply and demand for programs and projects.

Portfolio Management Standard, p. 18

Task 8 in the ECO in Performance

50. d.   Document communications required for successful Implementation

Portfolio management is a major culture change if set up and managed effectively. As a first step, a portfolio strategic plan is prepared, with a section in it involving communications required for successful change and implementation.

Portfolio Management Standard, p. 46

Task 1 in the ECO in Strategic Alignment

51. c.   Your focus is on events

Dashboards are status display mechanisms to monitor operational performance as they measure performance against targets and thresholds using right-time data. At the portfolio level, a strategic dashboard monitors execution of strategic objectives, and the purpose is to align around strategic objectives, emphasizing management.

Kerzner (2011), pp. 37–39

Portfolio Management Standard, p. 117

Task 3 in the ECO in Performance

52. c.   Stakeholder engagement from the start

This situation shows the importance of innovative communications management and engaging stakeholders from the start to best embrace portfolio management and support the transformation initiative.

PMI (2013) Change Management Guide, p. 59

Task 2 in the ECO in Communications

53. b.   Continuity and alignment of expectations are critical to success

Stakeholder analysis is necessary in managing strategic change since it helps to ensure continuity and also to align stakeholder expectations with the strategic changes and portfolio realignment that will result.

Portfolio Management Standard, p. 54

Task 3 in the ECO in Strategic Alignment

54. c. It has measures and templates for reporting

   The PMIS can build on historical information as it can contain previously collected metrics and the templates used for reporting. The portfolio manager then has the option of using what has been proven to be successful in the past or can make changes to foster continuous improvement.

   Portfolio Management Standard, p. 90

   Task 2 in the ECO in Governance

55. d. An emphasis on risk management may lead to new components

   Risk management influences the portfolio in many ways. In this situation, hurricanes are a risk that is a threat to the country. It then leads to adding new components as a way to mitigate this risk in the future.

   Portfolio Management Standard, p. 119

   Task 1 in the ECO in Risk Management

56. c. You need approval to update the portfolio strategic plan

   If the portfolio manager cannot implement the portfolio, he or she should recommend or gain approval to update the portfolio strategic plan

   Portfolio Management Standard, p. 62

   Task 7 in the ECO in Strategic Alignment

57. d. Have the Portfolio Review Board obtain approval by the executive team

   Approval by the executive team demonstrates to the rest of the organization the importance of collecting these metrics and targets to assess whether the portfolio is progressing as expected.

   Portfolio Management Standard, p. 91

   Task 3 in the ECO in Performance

58. a. Preparing the governance model

   The governance model defines the portfolio's decision-making process, rights and authorities, responsibilities, rules, and protocols.

   Portfolio Management Standard, p. 62

   Task 1 in the ECO in Governance

59. c. Can build forecasts based on trends in the data

This technique has many uses in the Manage Portfolio Information process. Data may be collected in a raw form that lacks complete context and may be from external sources especially for purposes of comparative analysis. Through this technique, data can be analyzed to ensure it has value. To make the data more meaningful, they can be used for forecasts and trends.

Portfolio Management Standard, p. 117

Task 5 in the ECO in Communications

60. c. Update existing resource portfolio allocations and schedules

As a result of supply and demand analysis, it often is necessary to then recommend to the Portfolio Governance Group that the portfolio be updated to change the component mix as required.

Portfolio Management Standard, p. 96

Task 6 in the ECO in Strategic Alignment

61. a. Structural risk

A structural risk is one that arises from the wider environment of the organization. It is similar to a systems risk. It is one that affects the overall portfolio, as in this situation, if it is not mitigated, the company's image and market share would be affected adversely.

Portfolio Management Standard, p. 122

Task 3 in the ECO in Risk Management

62. d. Legal

In this scenario, it is appropriate for representatives from the legal department to be part of phase-gate reviews as these products will require approval from regulatory and legislative bodies at various levels.

Portfolio Management Standard, p. 22

Task 1 in the ECO in Governance

63. d. Burn-down chart

    Such a report is helpful to compare planned, completed, or remaining work and is especially helpful when a baseline is set. A variety of resource reports are useful, as an output of the Manage Supply and Demand process, to indicate whether resource capacity is matched optimally against resource demand.

    Portfolio Management Standard, p. 96

    Task 7 in the ECO in Performance

64. a. Ensure redundant information is not provided

    While some redundancy cannot be avoided and may even be intentional to reach multiple recipients, the purpose of the reviews is to assess requirements to eliminate redundant information.

    Portfolio Management Standard, p. 112

    Task 4 in the ECO in Communications

65. d. Ensures components in each category have a common goal

    When components are in categories, they have a common goal and can be measured on the same basis regardless of where they are in the organization.

    Portfolio Management Standard, p. 64

    Task 3 in the ECO in Strategic Alignment

66. c. Sustainability

    While all the answers are possible metrics to collect, sustainability has become an area of greater interest in the past few years.

    Portfolio Management Standard, pp. 85–86

    Task 3 in the ECO in Performance

67. d. The portfolio management plan

    These dependencies and the influence of the components influence the approaches defined in the management plan for communications and risk management.

    Portfolio Management Standard, p. 60

    Task 4 in the ECO in Governance

68. b. There are external environmental factors to consider

In preparing the portfolio strategic plan external environmental factors need to be reviewed; in this situation, an example is the regulations affecting this industry, interest by external stakeholders such as consumer interest groups, marketplace conditions, and the fact that employees may be members of unions.

Portfolio Management Standard, pp. 38, 44

Task 4 in the ECO in Strategic Alignment

69. c. The roadmap is useful to show the portfolio's structure

The roadmap is an input to the Plan Communication Management Plan process as it shows the structure, interdependencies, and how components interrelate in the portfolio to best achieve organizational goals and objectives. Therefore consistent and timely communications of the roadmap and changes to it are part of portfolio communications especially with a transparent process.

Portfolio Management Standard, p. 108

Task 2 in the ECO in Communications

70. b. Reallocates resources

If some components are removed from the portfolio especially when their contribution to benefits is low, resources then are reallocated to components with a higher priority.

Portfolio Management Standard, p. 71

Task 6 in the ECO in Performance

71. a. Guide talent development

The priority of each portfolio component guides resource planning, hiring decisions, schedules, and capability allocations, which includes long-range talent development.

Portfolio Management Standard, p. 9

Task 3 in the ECO in Strategic Alignment

72. d. Maintain them on a watch list

  While it is doubtful these risks will ever occur, organizational strategies may change. For example the organization may acquire another company that does work in a regulatory environment. These low probability/low impact risks should be maintained on a watch list with trigger conditions set up to indicate that if they may occur as further action then would be needed.

  Portfolio Management Standard, p. 127

  Task 2 in the ECO in Risk Management

73. c. Can model scenarios for resource use based on priorities

  With scenario analysis, a tool and technique in the Manage Supply and Demand process, tools in resource management are available. They can model scenarios to best use resources to meet the portfolio requirements.

  Portfolio Management Standard, p. 95

  Task 7 in the ECO in Performance

74. a. Is documented in the portfolio strategic plan

  Prioritization models may be simple or complex. Many organizations use scorecards, but whatever approach is used, it is documented in the portfolio strategic plan.

  Portfolio Management Standard, p. 64

  Task 2 in the ECO in Strategic Alignment

75. a. Establish a portfolio management information system

  The portfolio management information system is a tool and technique in Authorize Portfolio process and indicates which portfolio components have assigned resources.

  Portfolio Management Standard, p. 79

  Task 2 in the ECO in Governance

76. a. Set it up to treat the project as an option

Options analysis is appropriate in this situation as the project is large, with risks, and requires a large capital investment. NPV is not appropriate as it is not a routine project. This approach can be extensive as different options are determined, but by doing so a potentially profitable project may become affordable when the risks may seem too great.

Milosevic, pp. 61–63

Portfolio Management Standard, p. 74

Task 7 in the ECO in Performance

77. a. Document this approach in the methodology section

The portfolio risk management plan should include a methodology section that includes the approach, tools, and data sources that will be used.

Portfolio Management Standard, p. 128

Task 2 in the ECO in Risk Management

78. d. Create the portfolio mix with the greatest potential

As the portfolio is optimized it evaluates the selection criteria and ranks components. The purpose is to create the portfolio mix that has the greatest potential to support organizational strategy collectively.

Portfolio Management Standard, p. 71

Task 6 in the ECO in Performance

79. a. Components are focused on alignment to strategic objectives

The objective of portfolio management is to align the portfolio to strategic objectives and approve only those components that support strategic objectives. The portfolio then is re-examined if the strategy changes.

Portfolio Management Standard, p. 46

Task 1 in the ECO in Strategic Alignment

80. c. Portfolio Management Plan

This plan is an input to the Manage Supply and Demand process. It contains guidelines in a number of areas to report risks, communicate with the team, and to recommend changes in components such as those that are due to constraints on resources.

Portfolio Management Standard, p. 94

Task 4 in the ECO in Governance

81. d. Included real-time dashboards

Real-time dashboards often are in a PMIS, and they are used to provide triggers or warning systems of possible risks, issues, and other market concerns as soon as the event occurs to decision makers.

Portfolio Management Standard, p. 117

Task 5 in the ECO in Communications

82. a. Add a component to enhance brand image

Before proceeding into new markets, the brand image of the company after the credit card fraud breech must be enhanced following principles of scenario analysis.

Portfolio Management Standard, p. 95

Task 7 in the ECO in Performance

83. c. Criteria to see if the risks are identified consistently with the organization's risk strategy

Many organizations have a risk strategy in place at the top level. If so, the risk management plan updates standard criteria, which the risk team can use to see if the risks they have identified are done in a manner consistent with the organization's risk strategy.

Portfolio Management Standard, p. 128

Task 2 in the ECO in Risk Management

84. b. Ask the PMO to gather this information for you

The PMO can support portfolio management in a number of ways, one of which is to provide program and progress information and metrics reporting.

Portfolio Management Standard, p. 17

Task 2 in the ECO in Performance

85. b. Technology capabilities and capacities

This situation is one in which technology is continually changing; criteria to consider therefore are technology capabilities and capacities.

Portfolio Management Standard, pp. 66–67

Task 4 in the ECO in Governance

86. c.   Determine whether there are any needs that are required to best implement the change

A readiness assessment is useful for bridging the gap between the 'as is' and 'to be' state, and it also points out any needs not yet addressed but required for this change.

Portfolio Management Standard, p. 55

Task 7 in the ECO in Strategic Alignment

87. c.   Financial report

In managing resource supply and demand, finances are resources so a financial report is useful to show any updates in funding.

Portfolio Management Standard, p. 94

Task 3 in the ECO in Performance

88. c.   Internal resource allocation

Portfolio governance is a set of processes used to select and prioritize components and to allocate limited internal resources as it works to accomplish strategy and objectives in the organization.

Portfolio Management Standard, p. 62

Task 1 in the ECO in Governance

89. a.   Resource allocation

Resources are limited in each organization, and it is rare to spend one's time working only on one program or project. Reports on resource use are of interest to see when people are required to support specific activities on assigned programs and projects.

Portfolio Management Standard, p. 112

Task 4 in the ECO in Communications

90. c.   Portfolio performance reports

While a number of different reports may be helpful, performance reports are especially useful as they serve as indicators to manage risks. When the portfolio is under way, new risks may be introduced, and mitigation of other risks could lead to more effective performance.

Portfolio Management Standard, p. 131

Task 4 in the ECO in Risk Management

91. c. Risk ratings

    This situation represents a risk to the company as the new product line requires new technology, and the well-known existing product was terminated; risk reports require updates as an output of the Optimize Portfolio process.

    Portfolio Management Standard, p. 77

    Task 2 in the ECO in Performance

92. c. Stakeholder expectations

    The charter includes the key and major stakeholders as well as their expectations and requirements.

    Portfolio Management Standard, p. 49

    Task 2 in the ECO in Strategic Alignment

93. b. Assess the current state of portfolio management

    The first step in implementing portfolio management is to assess the current state of any processes that may exist in the organization. It can then lead as to what is necessary.

    Portfolio Management Standard, p. 23

    Task 1 in the ECO in Performance

94. b. Submit a proposal for additions to it in the areas of portfolio and program management books

    As this portfolio is not delivering its expected value, the best approach is to suggest some changes to ensure addition of future components can be added to support organizational strategy.

    Portfolio Management Standard, p. 22

    Task 3 in the ECO in Governance

95. c. Prepare a vision

    A vision for portfolio management is required to clarify the direction to proceed. It should be prepared in a way that reflects cultural values and conveys meaning to stakeholders.

    Portfolio Management Standard, pp. 9, 21

    Task 8 in the ECO in Strategic Alignment

96. b. Can tell when drum resources will be needed

Bottleneck resources also are known as drum resources. The critical chain approach uses buffers and buffer management to manage uncertainty as it considers the effect of resource allocation, optimization, and leveling.

PMBOK, p. 178

Portfolio Management Standard, p. 93

Task 7 in the ECO in Performance

97. c. Program B

Using NPV as a way to recommend components for optimization, a dollar one year from today is worth less than a dollar today. The more the future is discounted, or the higher discount rate, the less the NPV of the component. If the NPV is high, the component then is ranked high, leading to Program B.

Milosevic, pp. 42–44

Portfolio Management Standard, p. 74

Task 7 in the ECO in Performance

98. b. Processes to support change

At the functional manager level, he or she has risk concerns with product development, the organization's products and services, and processes required to best support change. In this situation, whenever a new product is released, the department must be able to support it.

Portfolio Management Standard, p. 122

Task 3 in the ECO in Risk Management

99. d. Workmanship standards

The other answers are portfolio process assets; workmanship standards are an example of an external environmental factor under a category of governmental and industry standards.

Portfolio Management Standard, p. 38

Task 2 in the ECO in Governance

100. b. There is a lack of transparency impacting the selection process

This example shows that the project management processes impact those at the portfolio even if the project process is not set up to provide transparency into the project's performance; thereby impacting the portfolio selection process.

Portfolio Management Standard, p. 25

Task 9 in the ECO in Performance

101. a. Shows portfolio communication dependencies

With the financial reports preceding the Portfolio Review Board meetings, the calendar then shows these dependencies so the Board members have the financial data to review before the meeting and can use the data for decision making.

Portfolio Management Standard, p. 118

Task 5 in the ECO in Communications

102. b. Scenario analysis

Scenario analysis, a tool and technique in the Optimize Portfolio process, is useful in this situation to show a variety of portfolio scenarios with different components and the current components to evaluate outcomes with different assumptions.

Portfolio Management Standard, p. 75

Task 7 in the ECO in Performance

103. c. Capacity analysis

Such an analysis is useful to assess the amount of work that can be performed. It reviews internal and external resource availability to help determine the portfolio structure.

Portfolio Management Standard, p. 48

Task 5 in the ECO in Strategic Alignment

104. b. Update the portfolio management plan

This plan is updated for several reasons; one of which is when there is a change or revision in the categories and classification of the portfolio components.

Portfolio Management Standard, p. 70

Task 4 in the ECO in Governance

105. b.   A leading indicator of future sales potential

The portfolio performance plan describes metrics to be collected to assess portfolio performance. The emphasis is to see if the portfolio is performing as planned. A pipeline of new products indicates the organization is continuing its strategic objectives of being the market leader.

Portfolio Management Standard, p. 91

Task 3 in the ECO in Performance

106. d.   Risk appetite

Risk appetite, or the degree of uncertainty the organization is willing to take in anticipation of a reward, is included in the risk measures section of the portfolio risk management plan. It also includes risk categories and criteria for probability and impact, the probability and impact matrix, and stakeholder attitudes toward risk.

Portfolio Management Standard, p. 128

Task 2 in the ECO in Risk Management

107. c.   Provides an overall view of the portfolio

Such an approach is helpful in terms of optimizing and balancing the portfolio as well as for funding allocation. It is also a method to use during portfolio reviews especially if a component is to be terminated for some reason.

Milosevic, pp. 80–81

Portfolio Management Standard, pp. 75–76

Task 7 in the ECO in Performance

108. d.   Management's intent to prioritize the work to meet the strategic objectives

The portfolio management plan describes the approach to identify, approve, procure, prioritize, balance, and manage the work in the portfolio to meet strategic objectives.

Portfolio Management Standard, p. 39

Task 4 in the ECO in Governance

109. c. Portfolio milestones, risks, cost and schedule

     The program sponsor is providing funding, resources, and high-level scope requirements and requires regular information on these items plus the portfolio's return on investment. The same information is needed for portfolio sponsors.

     Portfolio Management Standard, p. 111

     Task 3 in the ECO in Communications

110. d. Defined roles and responsibilities

     While a number of organizational enablers can assist in the implementation process, one that is required is to have an understanding of the defined roles and responsibilities of those staff members who will be actively involved in the portfolio management process.

     Portfolio Management Standard, p. 30

     Task 7 in the ECO in Performance

111. d. It will be revised several times

     As component information is collected, in order to ensure the necessary level of completeness, the portfolio manager may need to revise it several times.

     Portfolio Management Standard, p. 64

     Task 4 in the ECO in Strategic Alignment

112. c. You need to update the portfolio management plan

     The sponsors and governance model are noted in this plan; the other stakeholders who will require or benefit from information as to what is under way with the portfolio then are added to this plan.

     Portfolio Management Standard, p. 107

     Task 1 in the ECO in Communications

113. d. A defined change control structure

     Significant changes in the organization's environment may lead to a new strategic direction that impacts portfolio resources and components; these changes then rely on a defined change control structure.

     Portfolio Management Standard, p. 63

     Task 4 in the ECO in Governance

114. b. The portfolio communications and risk plans also require updates

Actions in one process in the portfolio affect other processes. This example represents a strategic change, and therefore, it may affect these other plans.

Portfolio Management Standard, p. 30

Task 7 in the ECO in Strategic Alignment

115. b. Organizational integrity

Everyone should be concerned about risks to the integrity of the organization. Once integrity is lost, it is difficult or even impossible to regain it. Other risks of interest at all levels include transparency and corruption.

Portfolio Management Standard, p. 122

Task 3 in the ECO in Risk Management

116. a. Used sensitivity analysis

Also known as 'what if' analysis, this approach uses a variety of scenarios to determine the effect of certain actions or decisions in terms of resource allocation and available capacity and capability to do existing work and to pursue new work.

Portfolio Management Standard, p. 90

Task 7 in the ECO in Performance

117. c. States stakeholder expectations and requirements

The information in the performance management plan is useful as it builds a framework for managing resources and to help generate value; recognition of stakeholder expectations and requirements is necessary.

Portfolio Management Standard, p. 88

Task 4 in the ECO in Governance

118. d. Consolidate the reports that are submitted

Component reports are an input to the Manage Portfolio Information process. The portfolio manager consolidates the reports, evaluates them for impact on portfolio performance, and determines stakeholders to receive the report.

Portfolio Management Standard, p. 116

Task 5 in the ECO in Communication

119. b. Dependencies

As the portfolio strategic plan is updated as an output from the Develop Project Charter process, it reflects changes in the structure of the portfolio. Updates may involve relationships, dependencies, and goals of the portfolio components.

Portfolio Management Standard, p. 49

Task 4 in the ECO in Strategic Alignment

120. d. Portfolio

As an input to the Manage Portfolio Information process, the portfolio must be reviewed as new stakeholders will have greater interest and communications requirements once their components are approved.

Portfolio Management Standard, p. 115

Task 5 in the ECO in Communications

121. a. The portfolio management plan may need revision

If there are changes in the portfolio oversight process, the portfolio management plan should be updated as needed.

Portfolio Management Standard, p. 84

Task 4 in the ECO in Governance

122. d. Program 2

In determining this choice, rank each column. Program 2 then has the highest priority of the six components listed.

Portfolio Management Standard, p. 69

Task 5 in the ECO in Strategic Alignment

123. c. Determine trends

The purpose of status and trend analyses is to compare current portfolio data and recent trends with recent changes in the portfolio. In this situation, this analysis would be helpful to determine how often these risks occurred and use the results to further assess their overall impact to the portfolio.

Portfolio Management Standard, p. 127

Task 2 in the ECO in Risk Management

124. d.   A SWOT analysis

A SWOT analysis helps determine the value of the portfolio components in the marketplace and how the components may affect or be affected by competitors.

Portfolio Management Standard, p. 75

Task 6 in the ECO in Performance

125. a.   Foundational

In this situation, the company has been in existence for 40 years. A component category of foundational is appropriate to ensure investments are made in the company's infrastructure so it continues to grow.

Portfolio Management Standard, p. 68

Task 4 in the ECO in Strategic Alignment

126. c.   Can consolidate communications at the portfolio level

The portfolio is an input to the Develop Portfolio Communications Plan process. With an understanding of portfolio components, it affects the communications approach to follow and does not lead to a misunderstanding of stakeholder communications needs.

Portfolio Management Standard, p. 108

Task 2 in the ECO in Communications

127. d.   Scenario analysis

It is an analytical method used as the project charter and structure are established to enable decision makers to evaluate the possible outcomes based on different assumptions.

Portfolio Management Standard, p. 47

Task 1 in the ECO in Governance

128. b.   Changes, risks, and issues

Portfolio component teams are a stakeholder group. Program, project, and operational managers expect notification of portfolio changes, risks, and issues as they may affect their specific component and may require them to take specific actions to ensure continued alignment with the organization's goals and objectives.

Portfolio Management Standard, p. 111

Task 3 in the ECO in Communications

129. d. Work with a company that is Level 3 certified as a subcontractor

One approach to handle a negative risk such as this one is to transfer it to another party. Being a subcontractor, if selected, continues the firm in this situation's positive relationship with the customer and shows it is focused on customer relationship management.

Portfolio Management Standard, p. 123

Task 5 in the ECO in Risk Management

130. b. An issue aging chart

An issue aging chart is a useful tool at the portfolio level as it can display in a graph the number of open issues on one axis and their age on the other axis to pinpoint ones that then require resolution even if another meeting is needed. The chart also can show the total number of issues, as well as those considered to be high priority versus low priority. Issues can be shown by type in categories. The objective is to identify trends and take action if it is not favorable and take action to reduce the majority of outstanding issues.

Milosevic, 2003, p. 490–491

Portfolio Management Standard, p. 37

Task 4 in the ECO in Performance

131. a. Ensure such allocations are reflected in the portfolio's strategic plan

The portfolio strategic plan, among other things, documents factors to be considered in the portfolio to aid decision makers in aligning, authorizing, and controlling the portfolio; this includes allocations of funds.

Portfolio Management Standard, p. 39

Task 6 in the ECO in Strategic Alignment

132. d. Execute portfolio management processes and apply resources to components

The charter is the key document for the portfolio manager's activities and also defines the portfolio structure.

Portfolio Management Standard, p. 47

Task 1 in the ECO in Performance

133. d. Is a portfolio process asset

Portfolio process assets are an input to the Provide Portfolio Oversight process. In this process, historical performance information, open issues, and governance decisions are particularly useful.

Portfolio Management Standard, p. 83

Task 2 in the ECO in Governance

134. b. Insight from your stakeholder analysis

The portfolio's uniqueness and insight from stakeholder analysis are used to determine the importance of each of these assets.

Portfolio Management Standard, p. 116

Task 5 in the ECO in Communications

135. a. References to corporate risk guidelines

The organization may have guidelines, policies, and procedures in place, which would be referenced in the plan as they define risk strategy, tolerance, and thresholds.

Portfolio Management Standard, p. 123

Task 2 in the ECO in Risk Management

136. d. Combines and details forecasts of ongoing supply and demand

Intellectual capital is a resource type. The resource schedule is displayed as a histogram, and it then is useful for forecasts and for showing existing supply and demand.

Portfolio Management Standard, p. 90

Task 7 in the ECO in Performance

137. b. The link between organizational strategies and portfolio management

The graphical nature of the roadmap serves as a way to help communicate why portfolio management is being done as the components then are shown in terms of organizational strategies.

Portfolio Management Standard, p. 50

Task 8 in the ECO in Strategic Alignment

138. a. Engage stakeholders through targeted messages

Portfolio reports are expected to change and are an output of the Manage Portfolio Information process. In doing so, it is necessary to provide proactive and targeted information to continually engage stakeholders and to make sure reports are delivered in a timely manner and in the desired format.

Portfolio Management Standard, p. 118

Task 5 in the ECO in Communications

139. b. Shows the highest score

This example shows a single criterion approach as a pair-wise comparison of different components with one another; the one with the highest rank then is selected.

Portfolio Management Standard, pp. 68–69

Task 5 in the ECO in Strategic Alignment

140. d. Vision statement

The organization's vision statement, along with its mission statement, are examples of organizational process assets useful in developing the portfolio risk management plan. Organizational process assets are an input to the Develop Portfolio Risk Management Plan process.

Portfolio Management Standard, p. 125

Task 2 in the ECO in Risk Management

141. c. Stakeholder relationships

Portfolio process assets useful in developing the portfolio charter include documentation on stakeholder relationships, scope, benefits, and the goals of the portfolio.

Portfolio Management Standard, p. 48

Task 1 in the ECO in Performance

142. a. Immature but evolving

Such an inventory takes time to develop and often requires progressive elaboration as details are gathered on components. In its early stages it is immature and may not be optimized.

Portfolio Management Standard, pp. 43–44

Task 4 in the ECO in Strategic Alignment

143. d.   Prepare the portfolio management plan

This plan needs to be aligned with the strategic plan, roadmap, and charter, and while the plans are prepared in an iterative way, the management plan tends to follow the preparation of the other three documents.

Portfolio Management Standard, p. 57

Task 4 in the ECO in Governance

144. d.   Use surveys

Surveys and questionnaires are an excellent elicitation technique to facilitate participation from a large group of stakeholders.

Portfolio Management Standard, p. 61

Task 4 in the ECO in Governance

145. d.   Preparing and following a transparent communications strategy

This strategy focuses on satisfying stakeholder information needs, and through transparency, it helps the portfolio manager establish effective working relationships with stakeholders.

Portfolio Management Standard, p. 105

Task 2 in the ECO in Communications

146. c.   Use the roadmap

The roadmap is a high-level prioritization of the mapping over time of the portfolio, and it forms the initial basis to establish dependencies in the portfolio and external to it so they can be tracked.

Portfolio Management Standard, p. 39

Task 8 in the ECO in Strategic Alignment

147. c.   Provide an approach to assess risks in proposed components

The Portfolio Review Board will use this plan since it contains how risks are assessed in proposed components and to determine whether it is worthwhile to justify investments in high risk components before they are approved.

Portfolio Management Standard, p. 123

Task 2 in the ECO in Risk Management

148. d. Resources are unlimited

     In a projectized structure, resources are considered to be unlimited and can be procured to meet demand.

     Portfolio Management Standard, p. 92

     Task 7 in the ECO in Performance

149. c. Implement organizational structure analysis

     This is a tool and technique to use as the portfolio management plan is developed as depending on the size of the organization a team or a single person may manage the portfolio, and executives or a portfolio board may make decisions.

     Portfolio Management Standard, p. 61

     Task 4 in the ECO in Governance

150. c. Set up training for staff and stakeholders

     Portfolio management is a major change initiative; therefore training in it and why it is important is required for the staff members who will have day-to-day responsibilities in its implementation and for key stakeholders involved in this process.

     Portfolio Management Standard, p. 24

     Task 5 in the ECO in Communications

151. c. Show where there are gaps in focus, investment, or alignment

     This situation is when strategic alignment analysis is appropriate to best validate what is being done against organizational strategy updates.

     Portfolio Management Standard, p. 44

     Task 3 in the ECO in Strategic Alignment

152. a. Ways to maintain a balanced portfolio

     As an output of the Optimize Portfolio process, updates to the portfolio management plan may be needed in terms of the optimizing approach, criteria, and any other information about maintaining a balanced portfolio.

     Portfolio Management Standard, p. 77

     Task 4 in the ECO in Governance

153. d. Meeting minutes

Updates to portfolio process assets are an output of the Manage Portfolio Information process. One example is meeting minutes, especially those involving the governance group as they should be retained as part of the portfolio artifacts.

Portfolio Management Standard, p. 118

Task 6 in the ECO in Communications

154. a. Positive or negative risk

Using the highly specialized training provider is positive if the provider delivers as specified in the contract, and the organization does not need to hire these specialized resources. If the provider does not perform as expected, it then is a negative risk, and an alternative supplier must be located or resources with the needed competencies hired.

Portfolio Management Standard, p. 122

Task 3 in the ECO in Risk Management

155. b. Descriptors and categories may need change

The portfolio strategic plan aligns the portfolio with organizational strategy. If the strategy changes, the existing portfolio should be validated against the changes as the key descriptors, categories, and organization of the portfolio components may require change.

Portfolio Management Standard, pp. 65–66

Task 1 in the ECO in Strategic Alignment

156. c. Benefits

The portfolio strategic plan includes the prioritization model, benefits, assumptions, constraints, dependencies, and risks.

Portfolio Management Standard, p. 55

Task 7 in the ECO in Strategic Alignment

157. b. Reports on resource availability should be useful

Resources must be allocated to authorized components. Reports that show the updated resource pool provide data on these resources and any resources reallocated from terminated or canceled components. Portfolio reports are an output of the Authorize Portfolio process.

Portfolio Management Standard, p. 80

Task 3 in the ECO in Governance

158. d. Portfolio charter

Among other things, the charter links the portfolio to organizational strategy and describes how it will deliver value to the organization.

Portfolio Management Standard, p. 49

Task 9 in the ECO in Performance

159. c. Component C

Scenario analysis uses probabilities as a decision analysis method. In this example, Component C has the highest EMV at $16,500 and would be recommended.

Rechenthin, p. 59

Portfolio Management Standard, p. 48

Task 5 in the ECO in Strategic Alignment

160. b. Assigns an owner to each risk

The risk register can be part of the portfolio risk management plan as part of the roles and responsibilities section. It identifies the owners and other team members for the activities in the plan and roles and responsibilities; one of which can be to own the risk until it is closed and no longer affects the portfolio.

Portfolio Management Standard, pp. 128, 131

Task 2 in the ECO in Risk Management

161. b. Ensure compliance with standards

One of the key activities in the Authorize Portfolio process is to ensure compliance with organizational standards, a typical responsibility under the Knowledge Management Officer's purview.

Portfolio Management Standard, p. 81

Task 1 in the ECO in Governance

162. b. Acquire an automated PMIS

One reason why portfolio reports will change, as an output of the Manage Portfolio Information process, is if there is an improvement in the speed and quality of the information. In this situation, acquiring an automated PMIS rather than using spreadsheets will meet this need.

Portfolio Management Standard, p. 118

Task 5 in the ECO in Communications

163. d. Allocation of funds to different types of initiatives and their contribution to strategy may require revision

The portfolio strategic plan is prepared and updated frequently; one section in it involves allocation of funds. If there are changes as in this situation this section in the plan probably will require revision.

Portfolio Management Standard, p. 46

Task 2 in the ECO in Strategic Alignment

164. b. Consider funding allocations

The portfolio is an input to the Portfolio Oversight process. It is used to consider component proposals, changes, and changes to funding allocations among components, whether proposed or in progress.

Portfolio Management Standard, p. 82

Task 1 in the ECO in Governance

165. a. Communications are built around this policy

This policy would be stated in the portfolio management plan, and it is an input to the Manage Portfolio Information process.

Portfolio Management Standard, p. 115

Task 5 in the ECO in Communications

166. b. Organizational strategy and objectives

They serve as the starting point for the portfolio strategic plan, and the plan may address the organizational strategy at the corporate, organizational unit, and functional department levels.

Portfolio Management Standard, p. 44

Task 1 in the ECO in Strategic Alignment

167. b. Determine an overall score for each component

While the prioritization model may be simple or complex, the purpose is to obtain an overall score and rank for each component.

Portfolio Management Standard, pp. 45–46

Task 6 in the ECO in Strategic Alignment

168. a. Update the portfolio management plan

    If roles and responsibilities change as a result of the Manage Portfolio Value process, the portfolio management plan requires revision.

    Portfolio Management Standard, p. 104

    Task 4 in the ECO in Governance

169. d. Provides a way to share or optimize scarce resources

    The Center of Excellence or PMO in this situation provides support to portfolio management through providing an effective way to share and optimize scarce or common resources.

    Portfolio Management Standard, p. 62

    Task 6 in the ECO in Performance

170. a. Use the 'as is' and 'to be' state and graphically display the components by application areas

    In a strategic alignment analysis a structure can be prepared based on a high-level time line showing organizational areas at different ends of a spectrum from the current situation to that of the organization's strategic vision.

    Portfolio Management Standard, pp. 44–45

    Task 3 in the ECO in Strategic Alignment

Appendix: Study Matrix Practice Tests

Study Matrix—Practice Test 1

Practice Test Question Number	Performance Domain	Study Notes
1	Portfolio Risk Management	
2	Strategic Alignment	
3	Communications Management	
4	Governance	
5	Governance	
6	Portfolio Performance	
7	Strategic Alignment	
8	Portfolio Performance	
9	Portfolio Risk Management	
10	Portfolio Risk Management	
11	Portfolio Performance	
12	Communications Management	
13	Portfolio Performance	
14	Strategic Alignment	
15	Communications Management	
16	Governance	
17	Portfolio Performance	
18	Portfolio Performance	
19	Governance	
20	Strategic Alignment	
21	Strategic Alignment	
22	Portfolio Risk Management	
23	Governance	
24	Portfolio Performance	
25	Strategic Alignment	
26	Portfolio Performance	
27	Communications Management	
28	Portfolio Performance	
29	Communications Management	
30	Governance	

continued

Practice Test Question Number	Performance Domain	Study Notes
31	Portfolio Performance	
32	Strategic Alignment	
33	Portfolio Risk Management	
34	Portfolio Performance	
35	Governance	
36	Communications Management	
37	Strategic Alignment	
38	Portfolio Performance	
39	Portfolio Performance	
40	Governance	
41	Strategic Alignment	
42	Governance	
43	Portfolio Performance	
44	Strategic Alignment	
45	Portfolio Performance	
46	Strategic Alignment	
47	Governance	
48	Portfolio Performance	
49	Portfolio Risk Management	
50	Governance	
51	Portfolio Risk Management	
52	Strategic Alignment	
53	Communications Management	
54	Communications Management	
55	Portfolio Performance	
56	Portfolio Performance	
57	Strategic Alignment	
58	Governance	
59	Portfolio Performance	
60	Portfolio Risk Management	
61	Strategic Alignment	
62	Communications Management	

Practice Test Question Number	Performance Domain	Study Notes
63	Governance	
64	Strategic Alignment	
65	Portfolio Risk Management	
66	Portfolio Performance	
67	Governance	
68	Strategic Alignment	
69	Portfolio Performance	
70	Portfolio Risk Management	
71	Portfolio Performance	
72	Strategic Alignment	
73	Portfolio Performance	
74	Governance	
75	Portfolio Performance	
76	Communications Management	
77	Strategic Alignment	
78	Portfolio Performance	
79	Governance	
80	Strategic Alignment	
81	Portfolio Risk Management	
82	Portfolio Performance	
83	Governance	
84	Strategic Alignment	
85	Communications Management	
86	Portfolio Performance	
87	Strategic Alignment	
88	Portfolio Risk Management	
89	Governance	
90	Strategic Alignment	
91	Portfolio Performance	
92	Portfolio Performance	
93	Communications Management	

continued

Practice Test Question Number	Performance Domain	Study Notes
94	Strategic Alignment	
95	Governance	
96	Portfolio Performance	
97	Strategic Alignment	
98	Portfolio Performance	
99	Portfolio Risk Management	
100	Governance	
101	Strategic Alignment	
102	Governance	
103	Communications Management	
104	Strategic Alignment	
105	Portfolio Performance	
106	Portfolio Risk Management	
107	Portfolio Performance	
108	Strategic Alignment	
109	Governance	
110	Communications Management	
111	Portfolio Performance	
112	Governance	
113	Strategic Alignment	
114	Portfolio Risk Management	
115	Portfolio Performance	
116	Communications Management	
117	Governance	
118	Portfolio Risk Management	
119	Strategic Alignment	
120	Portfolio Performance	
121	Communications Management	
122	Governance	
123	Portfolio Risk Management	
124	Portfolio Performance	
125	Communications Management	

Practice Test Question Number	Performance Domain	Study Notes
126	Strategic Alignment	
127	Governance	
128	Portfolio Risk Management	
129	Strategic Alignment	
130	Strategic Alignment	
131	Portfolio Performance	
132	Strategic Alignment	
133	Portfolio Risk Management	
134	Governance	
135	Strategic Alignment	
136	Communications Management	
137	Communications Management	
138	Portfolio Risk Management	
139	Strategic Alignment	
140	Strategic Alignment	
141	Portfolio Risk Management	
142	Governance	
143	Portfolio Performance	
144	Strategic Alignment	
145	Portfolio Risk Management	
146	Communications Management	
147	Strategic Alignment	
148	Governance	
149	Portfolio Risk Management	
150	Strategic Alignment	
151	Communications Management	
152	Strategic Alignment	
153	Portfolio Performance	
154	Portfolio Risk Management	
155	Strategic Alignment	
156	Governance	

continued

Practice Test Question Number	Performance Domain	Study Notes
157	Portfolio Performance	
158	Portfolio Risk Management	
159	Strategic Alignment	
160	Communications Management	
161	Governance	
162	Portfolio Performance	
163	Governance	
164	Communications Management	
165	Portfolio Performance	
166	Governance	
167	Strategy	
168	Communications Management	
169	Governance	
170	Communications Management	

Study Matrix—Practice Test 2

Practice Test Question Number	Performance Domain	Study Notes
1	Strategic Alignment	
2	Governance	
3	Portfolio Risk Management	
4	Portfolio Performance	
5	Communications Management	
6	Strategic Alignment	
7	Portfolio Performance	
8	Governance	
9	Portfolio Performance	
10	Portfolio Risk Management	
11	Strategic Alignment	
12	Portfolio Performance	
13	Communications Management	
14	Portfolio Risk Management	
15	Strategic Alignment	
16	Governance	
17	Governance	
18	Strategic Alignment	
19	Communications Management	
20	Portfolio Risk Management	
21	Portfolio Performance	
22	Strategic Alignment	
23	Portfolio Performance	
24	Portfolio Performance	
25	Strategic Alignment	
26	Portfolio Risk Management	
27	Governance	
28	Portfolio Performance	
29	Communications Management	
30	Strategic Alignment	

continued

Practice Test Question Number	Performance Domain	Study Notes
31	Portfolio Risk Management	
32	Portfolio Performance	
33	Strategic Alignment	
34	Governance	
35	Portfolio Performance	
36	Strategic Alignment	
37	Portfolio Performance	
38	Portfolio Risk Management	
39	Governance	
40	Portfolio Performance	
41	Strategic Alignment	
42	Communications Management	
43	Portfolio Risk Management	
44	Governance	
45	Strategic Alignment	
46	Portfolio Performance	
47	Governance	
48	Portfolio Risk Management	
49	Portfolio Performance	
50	Strategic Alignment	
51	Portfolio Performance	
52	Communications Management	
53	Strategic Alignment	
54	Governance	
55	Portfolio Risk Management	
56	Strategic Alignment	
57	Portfolio Performance	
58	Governance	
59	Communications Management	
60	Strategic Alignment	
61	Portfolio Risk Management	
62	Governance	

Practice Test Question Number	Performance Domain	Study Notes
63	Portfolio Performance	
64	Communications Management	
65	Strategic Alignment	
66	Portfolio Performance	
67	Governance	
68	Strategic Alignment	
69	Communications Management	
70	Portfolio Performance	
71	Strategic Alignment	
72	Portfolio Risk Management	
73	Portfolio Performance	
74	Strategic Alignment	
75	Governance	
76	Portfolio Performance	
77	Portfolio Risk Management	
78	Portfolio Performance	
79	Strategic Alignment	
80	Governance	
81	Communications Management	
82	Portfolio Performance	
83	Portfolio Risk Management	
84	Portfolio Performance	
85	Governance	
86	Strategic Alignment	
87	Portfolio Performance	
88	Governance	
89	Communications Management	
90	Portfolio Risk Management	
91	Portfolio Performance	
92	Strategic Alignment	
93	Portfolio Performance	

continued

Practice Test Question Number	Performance Domain	Study Notes
94	Governance	
95	Strategic Alignment	
96	Portfolio Performance	
97	Portfolio Performance	
98	Portfolio Risk Management	
99	Governance	
100	Portfolio Performance	
101	Communications Management	
102	Portfolio Performance	
103	Strategic Alignment	
104	Governance	
105	Portfolio Performance	
106	Portfolio Risk Management	
107	Portfolio Performance	
108	Governance	
109	Communications Management	
110	Portfolio Performance	
111	Strategic Alignment	
112	Communications Management	
113	Governance	
114	Strategic Alignment	
115	Portfolio Risk Management	
116	Portfolio Performance	
117	Governance	
118	Communications Management	
119	Strategic Alignment	
120	Communications Management	
121	Governance	
122	Strategic Alignment	
123	Portfolio Risk Management	
124	Portfolio Performance	
125	Strategic Alignment	

Practice Test Question Number	Performance Domain	Study Notes
126	Communications Management	
127	Governance	
128	Communications Management	
129	Portfolio Risk Management	
130	Portfolio Performance	
131	Strategic Alignment	
132	Portfolio Performance	
133	Governance	
134	Communications Management	
135	Portfolio Risk Management	
136	Portfolio Performance	
137	Strategic Alignment	
138	Communications Management	
139	Strategic Alignment	
140	Portfolio Risk Management	
141	Portfolio Performance	
142	Strategic Alignment	
143	Governance	
144	Governance	
145	Communications Management	
146	Strategic Alignment	
147	Portfolio Risk Management	
148	Portfolio Performance	
149	Governance	
150	Communications Management	
151	Strategic Alignment	
152	Governance	
153	Communications Management	
154	Portfolio Risk Management	
155	Strategic Alignment	
156	Strategic Alignment	

continued

Practice Test Question Number	Performance Domain	Study Notes
157	Governance	
158	Portfolio Performance	
159	Strategic Alignment	
160	Portfolio Risk Management	
161	Governance	
162	Communications Management	
163	Strategic Alignment	
164	Governance	
165	Communications Management	
166	Strategic Alignment	
167	Strategic Alignment	
168	Governance	
169	Portfolio Performance	
170	Strategic Alignment	

References

Kerzner, H. (2010). *Project management best practices Achieving global excellence.* Second Edition. Hoboken, NJ: John Wiley & Sons.

Kerzner, H. (2011). *Project management metrics, KPIs, and dashboards A guide to measuring monitoring project performance.* Hoboken, NJ: John Wiley & Sons.

Milosevic, D. J. (2003) *Project management toolbox Tools and techniques for the practicing project manager.* Hoboken, NJ: John Wiley & Sons.

Project Management Institute. (2008). *A guide to the project management body of knowledge (PMBOK® guide)* – Fourth edition. Newtown Square, PA: Project Management Institute.

Project Management Institute. (2013). *A guide to the project management body of knowledge (PMBOK® guide)* – Fifth edition. Newtown Square, PA: Project Management Institute.

Project Management Institute. (2013). *Managing change in organizations: A practice guide.* Newtown Square, PA: Project Management Institute.

Project Management Institute. (2013). *The standard for portfolio management*—Third edition. Newton Square, PA: Project Management Institute.

Project Management Institute. (2013). *Portfolio Management Professional (PfMP®) Examination Content Outline.* Newton Square, PA: Project management Institute.

Rechenthin, D. (2013). *Project intelligence.* Newtown Square, PA: Project Management Institute.

Turner, R., Huemann, M., Anbari, F., and Bredillet, C. (2010). *Perspectives on projects.* London: Routledge.

Printed and bound by CPI Group (UK) Ltd, Croydon, CR0 4YY

22/10/2024

01777634-0018